SpringerBriefs in Applied Sciences and Technology

SpringerBriefs present concise summaries of cutting-edge research and practical applications across a wide spectrum of fields. Featuring compact volumes of 50 to 125 pages, the series covers a range of content from professional to academic.

Typical publications can be:

- A timely report of state-of-the art methods
- An introduction to or a manual for the application of mathematical or computer techniques
- A bridge between new research results, as published in journal articles
- A snapshot of a hot or emerging topic
- An in-depth case study
- A presentation of core concepts that students must understand in order to make independent contributions

SpringerBriefs are characterized by fast, global electronic dissemination, standard publishing contracts, standardized manuscript preparation and formatting guidelines, and expedited production schedules.

On the one hand, **SpringerBriefs in Applied Sciences and Technology** are devoted to the publication of fundamentals and applications within the different classical engineering disciplines as well as in interdisciplinary fields that recently emerged between these areas. On the other hand, as the boundary separating fundamental research and applied technology is more and more dissolving, this series is particularly open to trans-disciplinary topics between fundamental science and engineering.

Indexed by EI-Compendex, SCOPUS and Springerlink.

More information about this series at https://link.springer.com/bookseries/8884

Tin-Chih Toly Chen · Yi-Chi Wang

Artificial Intelligence and Lean Manufacturing

 Springer

Tin-Chih Toly Chen ⓘ
Department of Industrial Engineering
and Management
National Yang Ming Chiao Tung University
Hsinchu, Taiwan

Yi-Chi Wang ⓘ
Department of Industrial Engineering
and Systems Management
Feng Chia University
Taichung, Taiwan

ISSN 2191-530X   ISSN 2191-5318 (electronic)
SpringerBriefs in Applied Sciences and Technology
ISBN 978-3-031-04582-0   ISBN 978-3-031-04583-7 (eBook)
https://doi.org/10.1007/978-3-031-04583-7

This Springer imprint is published by the registered company Springer Nature Switzerland AG
The registered company address is: Gewerbestrasse 11, 6330 Cham, Switzerland

Contents

Chapter 1
Basics in Lean Management

1.1 Introduction

Lean manufacturing, or **lean sigma**, originated in Japan and is a well-known tool for improving the competitiveness of manufacturers around the world. Lean manufacturing improves the planning, control, and management of a manufacturing system by using simple and effective tools such as kanbans, pacemaker, value stream mapping, 5s, just-in-time (JIT), standard operating procedures, load leveling, pull manufacturing, and others, as illustrated in Fig. 1.1. Common features of these tools are transparency, ease of understanding and communication, and ease of use. However, the philosophy of low volume and high diversity and pull production in lean manufacturing may not be suitable for all types of factories. Nonetheless, some lean management concepts and techniques are of reference value for all factories.

Toyota production system (TPS) is considered as the predecessor of lean manufacturing. TPS has been successfully applied to factories and supply chains around the world to shorten cycle times, regulate outputs, facilitate decision-making processes, reduce costs, and enhance worker safety [1, 2].

So far, the concepts and techniques of lean manufacturing have been applied to non-manufacturing fields, forming the concept of so-called "lean thinking", which aims to "do more with less" [3].

According to Sanders et al. [4], there are four success factors for lean manufacturing:

- Supplier relationship;
- Process and control;
- Human factors;
- Customer focus.

In the view of Melo et al. [5], human factors and ergonomics are also key considerations when planning a lean work environment. In fact, lean manufacturing environments are more likely to ensure worker health and safety.

© The Author(s), under exclusive license to Springer Nature Switzerland AG 2022
T.-C. T. Chen and Y.-C. Wang, *Artificial Intelligence and Lean Manufacturing*,
SpringerBriefs in Applied Sciences and Technology,
https://doi.org/10.1007/978-3-031-04583-7_1

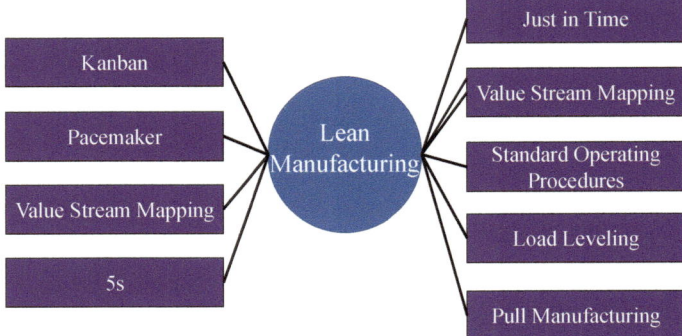

Fig. 1.1 Lean manufacturing technologies

1.2   Basic Concepts of Lean Manufacturing

1.2.1   3M and Seven Wastes

Lean manufacturing aims to eliminate three types of deviations [2, 6] that are illustrated in Fig. 1.2:

- **Muda**: Mudd includes activities that do not add value. The results of such activities are usually waste, i.e., the so-called seven types of wastes—overproduction, waiting, transportation, over processing, inventory, unnecessary motions, and product defects.
- **Mura**: Mura indicates the variability, inconsistency, unevenness, non-uniformity, or irregularity in production (in time, quantity, or quality). The existence of Mura leads to the **seven wastes**.
- **Muri**: Muri refers to situations in which operators or machines operate above their limits. Overburden, excessiveness, and unreasonableness are some synonyms of Muri. Muri may result from Mura or the excessive removal of Muda.

Therefore, activities are value-added if they avoid wastes and produce exactly what is needed, where and when it is needed.

Fig. 1.2 3M and seven wastes

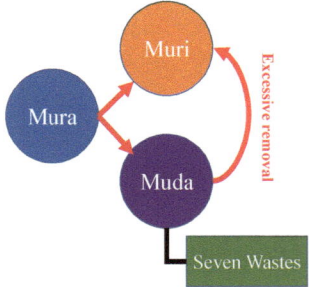

The elimination of 3M usually starts with eliminating Muda (i.e., seven wastes). Managers should immediately address overproduction and unnecessary waiting, shipping, handling, inventory, actions, or corrections. In fact, after the elimination of Muda, Mura, and Muri also decrease, which improves the working environment and also the working performance [5].

1.2.2   5S

5S include a series of shop floor improvement activities aimed at making the shop floor cleaner, tidier, more hygienic, and safer. Such activities are divided into five categories: Seiri, Seiton, Seiso, Seiketsu, and Shitsuke [7], as defined in Table 1.1. To the Japanese, 5S are actually a daily practice of life wisdom, so they are easy to be integrated into management practices [8]. 5S achieve cost-effectiveness by maximizing efficiency and effectiveness. In lean manufacturing, 5S are among the most prevalent and easily effective improvement activities. However, in many organizations, only the first three S activities were performed, which limited the possible benefits [9].

The rapid advancement of computer and information technologies has diversified the implementation of 5S activities. According to the findings of Gapp et al. [10], an organization website is a suitable channel for disseminating information about 5S practices. Whether **artificial intelligence (AI)**, as the most advanced computer and computing technology, can be applied to 5S activities is a topic of concern [11]. This is also a direction this book intends to explore.

1.2.3   Toyota Production System (TPS)

Toyota Production System (TPS) is the redesign of a mass production system [12]. It is a production method created and developed by Toyota by completely eliminating wastes to achieve good product quality, low costs, and short lead time (i.e., the time between a customer placing an order and the delivery of the order) [12]. It is a

Table 1.1 Definitions of 5S

Kaizen activity	Definition
Seiri	Throw away useless stuff
Seiton	Align, sort materials, workpieces, tools, finished goods, etc
Seiso	Clean the shop floor
Seiketsu	Keep the cleanliness of the shop floor
Shitsuke	Educate people to get used to the continuous implementation of 5S activities

concept and a law-abiding structure to improve the competitiveness of an enterprise. The current Toyota Production System is not in its completed form.

The two pillars of TPS are **Jidoka** (i.e., designing equipment that will automatically stop when an abnormality occurs and notify personnel of the abnormality) and **JIT**, both of which depend on a stable foundation. Methods to build this stable foundation include production leveling, standardizing work, and (continuous) improvement [6]. Major treatments taken in TPS include the conversion of linear production lines into U-shaped production cells, and the redesign of job shops into manufacturing cells. In addition, the production at each workstation runs on the **takt time** by making the processing time equal to or slightly less than the takt time and following a "make one, check one, move one" (MO-CO-MOO) procedure. For this purpose, single-cycle machines have built-in devices to inspect finished workpieces (poka-yokes) [13].

Although many people often equate TPS with lean management, the former is more inclined to applications in manufacturing, while the latter has been extended to applications in various industries (such as service industries, medical industries, and education industries).

1.2.4 Just in Time (JIT)

JIT is to provide the necessary amount of materials (or finished goods) where and when they are needed [14]. To this end, JIT uses pull production to control a manufacturing system. A workstation will deliver or produce a workpiece only when the next workstation needs it.

JIT is a management philosophy that encourages changes and improvements by reducing inventory. However, JIT is possible only when manufacturing systems have good product quality, high process reliability (or stability), low setup times, and low demand variability [15]. In particular, demand must be precisely predictable. Conversely, in order to realize JIT, a manufacturing system must solve problems such as poor product quality, easy machine failures, and unstable demand, thereby enhancing its competitiveness [16]. Other benefits of JIT include shorter lead times, improved ability to meet due dates, increased flexibility, easier automation, and better utilization of workers and equipment.

In pull methods such as JIT, the flow of information is tied to the flow of materials. In other words, valuable information, such as future demand variation, is not necessarily shared with everyone right away [14]. In contrast, for AI technologies, the real-time sharing of information is the basis for subsequent analyses. From a management perspective, it becomes a challenge to incorporate AI technology applications into a pull (or JIT) manufacturing system.

1.2.5 Total Productive Maintenance (TPM)

One of the keys to the success of lean manufacturing lies in the stability of the manufacturing system, especially the reliability of equipment [17]. In this regard, lean manufacturing has also developed its own set of knowledge, the so-called **total productive maintenance (TPM)** [18]. TPM is a philosophy that involves operators in maintaining their own equipment. TPM emphasizes that proactive and preventive maintenance provides the foundation for continued machine reliability and product quality [19].

TPM aims to improve productivity, quality costs, cost of products, delivery and movement of products, safety of operations, and morale of those involved (PQCDSM) [20], for which the following considerations are critical:

- Improvement actions should focus on parts that add value: Some motions of a machine do not add value and are considered as wastes that should be eliminated.
- Production at the necessary capacity: The goal is to generate no defects and produce according to the predetermined takt time [21]. No machine can process much shorter (faster) than the takt time
- A machine must be easy to maintain.
- The availability of a machine should be easily elevated.
- A machine has to act correctly when it should.
- The mold of a machine can be quickly changed.
- A machine can be easily relocated.
- Machines are the smaller, cheaper, or more labor-saving the better.

TPM is perhaps one of the lean manufacturing technologies that benefit most from the advancement of information and network technologies. For example, an operator who wants to maintain or even repair equipment by himself can obtain instructions from the equipment supplier through Internet messages or telex. Therefore, AI technologies are also expected to have a larger application space in TPM.

For example, Mohan et al. [21] considered the time to the next failure of a high-pressure hydraulic sand molding machine as a time series and applied an adaptive autoregressive integrated moving average (ARIMA) model [22] to predict it. After successfully predicting the time to the next failure, if the time is too close, the regular maintenance of the machine can be advanced. In some past studies, TPM problems have been formulated as semi-Markov decision processes (SMDPs), which were solved using dynamic programming (DP) when problem sizes were small. Encapera et al. [23] incorporated **reinforcement learning** techniques into dynamic programming, enabling it to solve large-scale TPM problems.

1.2.6 Kanbans

A **kanban** is a tool used by the next workstation to obtain WIP from the previous workstation. A kanban is some form of signals for a "please produce" or "please pick up" instruction in pull production, as a permission to produce or pick up [14]. These actions are performed according to the requirements of the next workstation. Therefore, kanbans are the shop floor control mechanism of a pull (or JIT) manufacturing system. Kanbans are not just used to pull workpieces, but also used to visualize and control the WIP in the factory.

Kanbans are divided into two categories based on their functionalities. "Production kanbans" require the previous workstation to manufacture workpieces, and "transportation kanbans" require movers to transport workpieces. "Transportation kanbans" can be further divided into those used in the factory and those used when delivering to customers. Each function may have a different form of kanbans.

Since kanbans are just a tool, it is unrealistic to expect a sudden drop in inventory or costs after the introduction of kanbans.

1.2.7 Spaghetti Diagram

A spaghetti diagram is a visual tool that uses a continuous flow to trace the path of an item (or activity) through a system (or process) [24]. A spaghetti diagram is a process analysis tool. It enables analysts to identify redundancies in workflows and opportunities to expedite process flows. Video-based time analysis tools are often used to help draw the spaghetti diagram of a manufacturing system. A spaghetti diagram is an application of the facility layout of a manufacturing system by drawing the flows of people (or robots) onto it with curves. After the spaghetti diagram is drawn, the frequency and total length of movements can be estimated and then reduced. Another equally important goal is to reduce the likelihood that movements will conflict with each other. After taking effective treatments to achieve these goals, a new spaghetti diagram was drawn.

Some existing mathematical or statistical software can be used to draw a spaghetti diagram. For example, Daneshjo et al. [25] embedded the facility layout of the manufacturing system as a figure into an excel worksheet, and then applies excel built-in functions to facilitate the calculations required for subsequent analyses.

The analysis results of a spaghetti diagram are also valuable for the implementation of 5S activities. For example, a very intuitive consideration is that there should be no accumulation of materials (or garbage) or greasy ground where movements highly overlap.

1.2.8 Drum Buffer Rope (DBR)

Similar to Kanban, DBR also aims to control WIP in manufacturing systems. The drum is the bottleneck of a manufacturing system, limiting the output rate of the manufacturing system. A buffer is the stock built to keep the drum from starving. When the stock in the buffer is about to run out, the rope pulls the production at upstream workstations to build up more stock. DBR is a production scheduling or control method developed based on the theory of constraints (TOC) [26]. The philosophy of TOC is that complex systems imply simplification. No matter how complex a system is at any time, there are very few variables in the system or only one, called the constraint, that will limit (or hinder) the system from reaching higher goals.

For a larger or more complex manufacturing system, full pull production is extremely difficult. In this case, DBR provides a simple and effective way to carry out pull production. In fact, the idea of avoiding the waste of bottleneck machine capacity in DBR can also be applied to non-pull production systems. One difficulty DBR encounters is that the bottleneck machines in many production systems may change, which makes it difficult for the DBR mechanism to perform a long-term, stable production control. To address this issue, various methods have been proposed in the literature. For example, Zhang et al. [27] proposed the dynamic bottleneck detection (DBD) approach, in which different heuristics were applied to schedule jobs that will go to bottleneck and non-bottleneck workstations, as summarized in Table 1.2. Chen [28] mentioned the disadvantages of the DBD approach:

- Many factories have more than two job priorities, such as "normal", "hot", "super hot", etc. Therefore, jobs may have to be divided into more categories.
- The formula used to assess the extent to which a machine is a bottleneck contains many parameters. How the values of these parameters are determined (or optimized) is an issue.
- Jobs with similar working conditions may be assigned to different categories.

Table 1.2 Job categories and scheduling heuristics in the DBD approach

Job category	Conditions	Scheduling heuristic
First priority	• Hot jobs	Critical ratio (CR) + first in firs out (FIFO)
Second priority	• Not hot jobs • The next workstation is a weak bottleneck	Shortest processing time until the next bottleneck (SPNB) + CR + FIFO
Third priority	• Not hot jobs • The next workstation is a strong bottleneck	Shortest processing time (SPT) + CR + FIFO
Fourth priority	• None of the above	CR + FIFO

1.2.9   Value Stream Map (VSM)

Value Stream Mapping (VSM) is a lean manufacturing (or lean enterprise management) technique used to document, analyze, and improve the flow of information or materials needed to manufacture a product or provide a service to a customer [29]. It is used to analyze the various actions in designing, selling, and manufacturing a product (or providing a service) and classifying them into three categories to further eliminate wastes:

- Actions that can create value that customers can feel.
- Wastes that do not added value but cannot be ruled out at present.
- Redundant steps that are of no value to customers and can be removed immediately.

A VSM is basically composed of six parts: customer, production planning and control, suppliers, receiving, production steps, and delivery, as illustrated in Fig. 1.3. The focus of improvement is usually on how to shorten the cycle time of the manufacturing process. Common treatments include applying pull production, adjusting the capacity of the storage space, and improving the operation at each workstation to shorten the processing time, and splitting or merging operations. A VSM is drawn before and after improvement.

While 5S is the most widely used lean manufacturing technology, value stream mapping is considered one of the most effective lean manufacturing technologies.

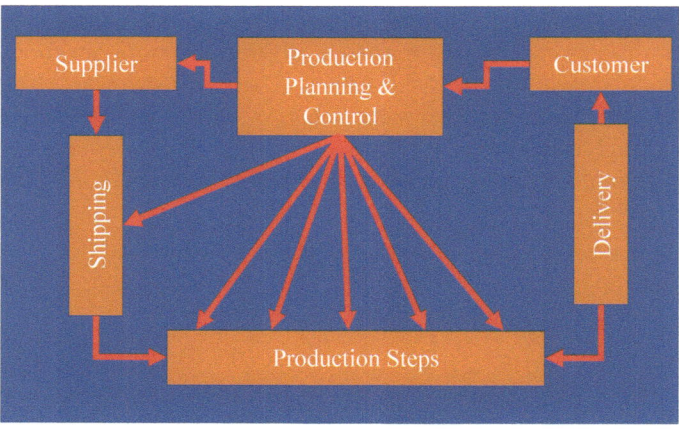

Fig. 1.3 Structure of a VSM

1.3 Evolution of Lean Manufacturing

Although lean manufacturing is a fairly old concept, it is still quite useful in the face of a changing international environment: intensified market competition, highly customized products, and shortened product life cycles [8]. However, with the widespread application of computer, sensor, communication, and AI technologies, the way to creating value has changed. For example,

- It is now possible for an upstream manufacturer to receive orders directly from end customers, enabling full pull production.
- The Internet of machines (IoM) is a vision where machines can communicate directly with each other, making pulling workpieces between them more immediate.
- The collection box of workpieces is attached with a radio frequency identification (RFID) tag, which can be used to track its flow and avoid long-term inventory in the same place.
- Telecommunication enables equipment operators to maintain and repair equipment on their own, under the direction of equipment suppliers, to realize TPM [30].
- Perico and Mattioli [30] defined lean 4.0 as the combination of lean manufacturing and Industry 4.0 to create intelligent networks along a value chain that can work separately and control each other autonomously.

However, for factory managers who are accustomed to lean production, the inscrutable AI is attractive but difficult to understand and accept. Applying AI to lean production is a more feasible way to combine the advantages of these two disciplines.

1.4 Organization of This Book

This book is intended to provide technical details on the applications of AI to lean manufacturing, including methodologies, system architectures, software and hardware, examples, and various applications. In addition to introducing traditional AI methods (such as fuzzy logic, artificial neural networks (ANN), and machine learning, RFID), this book will also mention some newer AI developments (including Internet of Things (IoT), IoM, edge computing, cloud computing, deep learning, big data analytics, etc.), and the applications of these AI technologies in the field of lean production.

In specific, the outline of the present book is structured as follows.

In the current chapter, lean manufacturing is first defined. Then, some basic concepts in lean manufacturing, such as 3M and seven wastes, TPM, JIT, TPS, kanbans, spaghetti diagram, DBR, and VSM are introduced. Although the terms mentioned in this chapter do not cover all lean manufacturing techniques, they are among the most representative. In addition, these lean manufacturing techniques can be applied not only in manufacturing, but also in other industries such as education, healthcare, services, and medical. The chapter concludes by referring to some trends in lean manufacturing theory and practices.

Chapter 2, Artificial Intelligence in Manufacturing, provides an introduction of some artificial intelligence (AI) methods and their applications in manufacturing. This chapter begins by defining AI. Then, existing AI methods are divided into several categories. Some representative applications of AI methods in each category are reviewed. Subsequently, the potential of AI technology applications in lean manufacturing is illustrated by referring to some instances retrieved from the literature and related reports. Some conclusions to date and issues facing lean manufacturing practitioners are also reported. Furthermore, a procedure for applying AI technologies to lean manufacturing systems is provided.

Chapter 3, AI Applications to Kaizen Management, which describes how to apply AI technologies to assist a series of kaizen (i.e., improvement) activities, such as 5S activities to improve the working environment, predictive maintenance activities to improve equipment reliability (or availability), cycle time reduction activities, and the evaluation of the outcome of all kaizen activities (i.e., leanness) in a manufacturing system.

Chapter 4, AI Applications to Pull Production, JIT and Production Leveling, involving three important topics in lean manufacturing, namely how to implement pull production in a manufacturing system, how to realize JIT, and how to balance the production capacity of a manufacturing system, especially when the production conditions are subject to uncertainty or the manufacturing system is complicated. Several AI technologies to address the three issues are introduced, including fuzzy arithmetic, backpropagation networks, fuzzy mathematical programming, and cloud computing.

Chapter 5, AI Applications to Shop Floor Management in Lean Manufacturing, first defines shop floor management. Then, some to shop floor management activities in lean manufacturing systems, such as lean data, lean maintenance, digitalized kanbans, and single minute exchange of die (SMED), are discussed. AI technologies that can be applied to assist in these activities include ANNs, edge computing (or edge intelligence), RFID, machine learning, Industry 4.0, and genetic programming. Some examples are also given to illustrate the application of these AI technologies.

References

1. A. Popa, R. Ramos, A.B. Cover, C.G. Popa, Integration of artificial intelligence and lean sigma for large field production optimization: application to Kern River Field, SPE Annual Technical Conference and Exhibition (2005)
2. K. Antosz, L. Pasko, A. Gola, The use of artificial intelligence methods to assess the effectiveness of lean maintenance concept implementation in manufacturing enterprises. Appl. Sci. **10**(21), 7922 (2020)
3. M. Poppendieck, Principles of lean thinking. IT Manag. Select **18**, 1–7 (2011)
4. A. Sanders, C. Elangeswaran, J.P. Wulfsberg, Industry 4.0 implies lean manufacturing: research activities in industry 4.0 function as enablers for lean manufacturing. J. Ind. Eng. Manag. **9**(3), 811–833 (2016)
5. T. Melo, A.C. Alves, I. Lopes, A. Colim, Reducing 3M by improved layouts and ergonomic intervention in a lean journey in a cork company, in *Occupational and Environmental Safety and Health II* (2020), pp. 537–545
6. N. Toshiko, *Kaizen Express* (Lean Enterprise Institute, 2009)
7. J. Michalska, D. Szewieczek, The 5S methodology as a tool for improving the organization. J. Achievements Mater. Manuf. Eng. **24**(2), 211–214 (2007)
8. T. Osada, *5S—Handmade Management Method* (JIPM, 1989)
9. C.D. Chapman, Clean house with lean 5S. Qual. Prog. **38**(6), 27–32 (2005)
10. R. Gapp, R. Fisher, K. Kobayashi, Implementing 5S within a Japanese context: an integrated management system. Manag. Decis. **46**(4), 565–579 (2008)
11. B. Kassem, F. Costa, A.P. Staudacher, Discovering artificial intelligence implementation and insights for lean production, in *European Lean Educator Conference* (2021), pp. 172–181
12. J.T. Black, Design rules for implementing the Toyota production system. Int. J. Prod. Res. **45**(16), 3639–3664 (2007)
13. M. Saruta, Toyota production systems: the 'Toyota way' and labour–management relations. Asian Bus. Manag. **5**(4), 487–506 (2006)
14. M.L. Junior, M. Godinho Filho, Variations of the kanban system: literature review and classification. Int. J. Prod. Econ. **125**(1), 13–21 (2010)
15. H. Groenevelt, The just-in-time system. Handbooks Oper. Res. Manag. Sci. **4**, 629–670 (1993)
16. T. Chen, Creating a just-in-time location-aware service using fuzzy logic. Appl. Spat. Anal. Policy **26**(9), 287–307 (2016)
17. H. Pačaiová, G. Ižaríková, Base principles and practices for implementation of total productive maintenance in automotive industry. Qual. Innov. Prosperity **23**(1), 45–59 (2019)
18. leanproduction.com, TPM (total productive maintenance) (2021). https://www.leanproduction.com/tpm/
19. C.J. Bamber, J.M. Sharp, M.T. Hides, Factors affecting successful implementation of total productive maintenance: a UK manufacturing case study perspective. J. Qual. Maintenance Eng. **5**(3), 162–181 (1999)
20. R.M. Ali, A.M. Deif, Dynamic lean assessment for takt time implementation. Procedia CIRP **17**, 577–581 (2014)
21. T.R. Mohan, J.P. Roselyn, R.A. Uthra, D. Devaraj, K. Umachandran, Intelligent machine learning based total productive maintenance approach for achieving zero downtime in industrial machinery. Comput. Ind. Eng. **157**, 107267 (2021)
22. H. Nakayama, S. Ata, I. Oka, Predicting time series of individual trends with resolution adaptive ARIMA, in *2013 IEEE International Workshop on Measurements & Networking* (2013), pp. 143–148
23. A. Encapera, A. Gosavi, S.L. Murray, Total productive maintenance of make-to-stock production-inventory systems via artificial-intelligence-based iSMART. Int. J. Syst. Sci. Operat. Logist. **8**(2), 154–166 (2021)
24. K. Senderská, A. Mareš, Š Václav, Spaghetti diagram application for workers' movement analysis. UPB Sci. Bull. Ser. D Mech. Eng. **79**(1), 139–150 (2017)

25. N. Daneshjo, V. Rudy, P. Malega, P. Krnáčová, Application of Spaghetti diagram in layout evaluation process: a case study. Technol. Edu. Manag. Inf. J. **10**(2), 573–582 (2021)
26. J.F. Cox III, J.G. Schleier Jr., *Theory of Constraints Handbook* (McGraw-Hill Education, 2010)
27. H. Zhang, Z. Jiang, C. Guo, Simulation-based optimization of dispatching rules for semiconductor wafer fabrication system scheduling by the response surface methodology. Int. J. Adv. Manuf. Technol. **41**(1–2), 110–121 (2009)
28. T. Chen, A fuzzy-neural DBD approach for job scheduling in a wafer fabrication factory. Int. J. Innov. Comput. Inf. Control **8**(6), 4024–4044 (2012)
29. M. Braglia, G. Carmignani, F. Zammori, A new value stream mapping approach for complex production systems. Int. J. Prod. Res. **44**(18–19), 3929–3952 (2006)
30. P. Perico, J. Mattioli, Empowering process and control in lean 4.0 with artificial intelligence, in *Third International Conference on Artificial Intelligence for Industries* (2020), pp. 6–9

Chapter 2
Artificial Intelligence in Manufacturing

2.1 Artificial Intelligence (AI)

The definition of artificial intelligence (AI) is undetermined. With the evolution of computer, network, and sensor technologies, the meaning of AI will still change [1]. Perico and Mattioli [2] divided AI technologies into two categories:

- **Data-driven AI** (i.e., brain-style learning), including artificial neural networks, machine learning (supervised learning, unsupervised learning, statistical learning), evolutionary computing, fuzzy logic, etc. Data-driven AI is often used in the context of pattern recognition, classification, clustering, or perception.
- **Symbolic AI** (i.e., modeling and knowledge reasoning), including ontology, semantic graphs, knowledge-based systems, reasoning, etc. Multi-criteria decision-making, production planning, and job scheduling are typical applications of this category.

In the view of Pandl et al. [1], AI enables computers to execute tasks that are easy for people to perform but difficult to describe formally. This seems to conflict with the philosophy of lean management, in which manufacturing systems are managed in a way that is transparent (i.e., easy to understand and communicate). Therefore, combining AI with lean management is a challenge. However, in the face of increasingly intense industrial competition, neither AI nor lean management alone is enough to enhance the competitiveness of an enterprise. A way to address this challenge is to apply the so-called explainable AI (XAI) [3].

AI technologies have been widely used in manufacturing. So far, the application of AI in manufacturing has brought closer connections between people, machines, and information technologies, enabling manufacturers to better optimize processes and predict problems [4]. However, AI technologies applied in manufacturing are naturally different from those applied in other fields. In a manufacturing environment, AI must help people, machines, and systems communicate with each other. In contrast,

© The Author(s), under exclusive license to Springer Nature Switzerland AG 2022
T.-C. T. Chen and Y.-C. Wang, *Artificial Intelligence and Lean Manufacturing*,
SpringerBriefs in Applied Sciences and Technology,
https://doi.org/10.1007/978-3-031-04583-7_2

in other fields, AI technologies are mostly applied to assist people. Nevertheless, service-oriented manufacturing [5] may eliminate this difference.

In the future, there is a trend toward merging AI with Internet technologies [5]. There are also researchers who equate Industry 4.0 with AI [4]. Some researchers have proposed the concept of AI 2.0 [6–8], but the definitions given by them were not consistent. The main features of AI 2.0 include deep learning, Internet-based AI, augmented intelligence, cross-media reasoning, and others [9].

Among existing AI technologies, machine intelligence is a field of particular interest, as introduced below.

2.1.1 Machine Intelligence

Machine learning is a mainstream of AI. **Machine learning** is a form of data analysis that uses algorithms to continuously learn from data [10]. Machine learning allows computers to recognize hidden patterns without actual programming. The key aspect of machine learning is that when models are exposed to new data sets, they adapt to generate reliable and consistent outputs. There are four categories of machine learning:

- **Supervised machine learning**: Supervised machine learning employs a training dataset to teach a model to predict output variables from input variables. This training dataset includes inputs and correct outputs (i.e., actual values). The algorithm uses a loss function to measure its accuracy, adjusting until the error is sufficiently minimized [11].
- **Unsupervised machine learning**: In unsupervised machine learning, training dataset is not employed to teach a model. Instead, a model itself finds the hidden patterns and insights from a given dataset [12]. Machine learning has proven to be very efficient at classifying images and other unstructured data, which is very difficult to handle using classic rule-based software [13].
- **Semi-supervised machine learning**: Semi-supervised learning combines clustering (unsupervised learning) and classification (supervised learning) algorithms. First, an unsupervised machine learning technique is applied to cluster objects based on their similarities. These clusters are labeled and then used to classify new objects [13].
- **Reinforcement machine learning**: In reinforcement machine learning, an agent makes a series of decisions to achieve a goal, such as using trial and error to come up with solutions to a problem, in an uncertain, potentially complex environment. The agent is rewarded or punished for the action it takes. Its goal is to maximize the total (or cumulative) reward [14]. Reinforcement machine learning is similar to the concept of dynamic programming in operations research.

Wuest et al. [15] provided an overview of machine learning techniques and described their successful applications in a manufacturing environment. The most commonly applied machine learning techniques included.

- **Inductive learning**, such as decision tree induction and rule induction.
- **Instance-based learning**, such as case-based reasoning.
- **Genetic algorithms (GAs)**: GAs treat feasible solutions to a problem as genes. These feasible solutions are then improved in a manner similar to genetic evolution [16].
- **Artificial neural networks (ANNs)** (for supervised, unsupervised, or reinforcement machine learning): An ANN is a network architecture that mimics the connections of biological neurons, in which artificial neurons receive, process, and send signals to others to perform a potentially complex task of classification, reasoning, or prediction [17].
- **Bayesian approaches**: Bayesian inference is a statistical inference method. First, inferences, classifications, or predictions are made based on certain assumptions. Previous assumptions are updated as new evidence or information becomes available [18].

Some examples of AI applications in manufacturing are given in the following sections.

2.2 AI Applications in Manufacturing

2.2.1 Inductive Learning

The cycle time, flow time, or manufacturing lead time of a job is the time required for the job to go through the factory. Estimating and shortening the cycle times of jobs is crucial for enhancing the competitiveness of the company. To this end, Wu and Chen [19] proposed a hybrid classification and regression tree (CART)-backpropagation network (BPN) approach to estimate job cycle times in a wafer fabrication factory (wafer fab), in which jobs were classified using a CART before cycle times were estimated using BPNs. Two machine learning techniques, decision tree induction, and ANNs, were applied in this application.

3D printing is a key development in additive manufacturing. Wang et al. [20] mentioned some of the difficulties faced by 3D printing researchers and practitioners, including high entry barriers of design for additive manufacturing (DfAM), limited material library, various defects, and inconsistent product quality. They believed that machine learning can help overcome these difficulties. For example, in designing metamaterials, machine learning techniques were applied to predict the properties of a metamaterial. In addition, hierarchical clustering and support vector machines were used jointly to optimize the material distribution of a 3D printed structure within a given design space subject to specific loads and constraints.

The fluctuation smoothing rule for cycle time variation (FSVCT) is an effective rule for scheduling jobs in a wafer fab to reduce the variation of job cycle times [21]. However, its rule content is fixed for different wafer fabs. If the rule content could be tailored to a specific wafer fab, it would be more effective. For this purpose, Chen et al. [22] first normalized the variables in the traditional FSVCT rule to propose the nonlinear FSVCT rule. Subsequently, two parameters were added to the nonlinear FSVCT rule to tailor the rule to a specific wafer fab.

Selecting suitable suppliers is a critical task to forming a sustainable supply chain. To this end, Amindoust et al. [23] established a fuzzy inference system (FIS) consisting of many smaller FISs to evaluate the performances of a candidate supplier along various dimensions. In theory, the fuzzy inference rules in a FIS can be derived by mining historical data or based on experts' subjective experiences. Amindoust et al. adopted the latter method.

2.2.2 Instance-Based Learning

A popular instance-based learning method is case-based reasoning (CBR). In Madhusudan et al. [24], the representative workflow designs of manufacturing systems were considered as cases. They discussed how to store, retrieve, reuse, and combine these cases. In designing the workflow of a new manufacturing system, the similarity between the new manufacturing system and existing cases was compared. Then, the workflow design of the new manufacturing system was generated from those of existing cases using the weighted average (WA) method.

In order to save the energy consumption of air-conditioning operations in the office space, González-Briones et al. [25] applied CBR. First, sensors were installed in the office space to detect how many people were in the office at different days and times. Then, the collected data were analyzed to build cases for predicting the number of people in the office. In different cases, air conditioners operated in different ways to save energy consumption.

Scheduling jobs in a semiconductor manufacturing factory is an extremely complicated task, as hundreds of machines and tens of thousands of jobs are involved. As a result, the production conditions in a semiconductor manufacturing system are constantly changing and are difficult to be modeled as cases. To overcome this difficulty, Jim et al. [26] modeled the production conditions in a semiconductor manufacturing system as a Petri net, and then mapped the states of the Petri net to cases. Subsequently, CBR could be applied to generate a scheduling plan for current production conditions.

The internal due date of an order is the time when the order can be completed and delivered to the customer, which is usually assessed internally by the factory production planner. Therefore, an internal due date should be as early as possible to be attractive. However, it depends on the flow times of all jobs of the order. To determine the internal due date for an order in a wafer fab, Chang et al. [27] proposed a CBR method in which a case was represented by a vector containing the six attributes of an order: the average queue length, processing time, the number of jobs in the factory, the number of jobs in the queue, factory load, and flow time. The flow time

Fig. 2.1 Procedure for implementing CBR

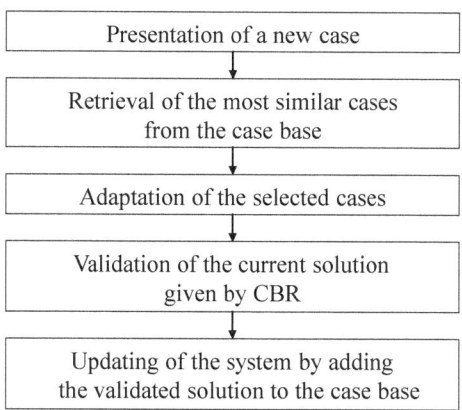

of a new order was estimated based on its first five attributes by comparing the new order with existing cases.

The measurement of the similarity between cases in CBR is valuable concept for single minute exchange of die (SMED) in lean manufacturing, in which products with similar setups should be processed consecutively. The similarity between two cases is usually inversely proportional to their Euclidean distance:

$$d(A, B) = \sqrt{\sum_{i=1}^{n} (A_i - B_i)^2} \qquad (2.1)$$

where A and B are two cases. A_i and B_i denote their values in the i-th dimension, respectively; $i = 1 \sim n$.

The procedure for implementing CBR is illustrated in Fig. 2.1.

2.2.3 Genetic Algorithms

A common practice in production scheduling is to group jobs with similar production conditions together to reduce setup times, in line with the philosophy of SMED in lean manufacturing [28]. Doing so also reduces processing times as operators become skilled in repeating similar production. In the job classification result, the variation between jobs in the same cluster is minimized, while differences between different clusters of jobs are maximized. This is undoubtedly an optimization problem. Both GAs and evolutionary computing can be applied to assist in the solution of this optimization problem. GA can also be applied to optimize the setting of parameters in an evolutionary computing method [29].

Chen and Lin [30] established a systematic procedure for comparing various applications of smart and automation technologies to ensure the long-term operation of a factory amid the COVID-19 pandemic. Alpha-cut operations (ACO) were applied to precisely derive the relative priority of a criterion for assessing a smart and automation technology application. However, ACO was time-consuming. To address this issue, an ACO problem was split into two constrained optimization subproblems. Then, GA was applied to help find the global optimal solution to each subproblem.

Kunyawan et al. [31] regarded a textile factory as a job shop, and then applied GA to solve the job scheduling problem of the textile factory. The objective function was to minimize the makespan. Problems of this type are usually NP-hard. In the proposed methodology, first, some scheduling plans were developed subjectively and then evaluated. Then, good schedules were retained as the first population of genes. GA was then applied to evolve the genes to make better scheduling plans.

Hong and Yo [32] formulated the energy management problem of a factory power system as a stochastic mixed integer-linear programming problem by considering the uncertainty in photovoltaic energies. The objective function was to minimize the expected value of the net costs of generating electricity from the micro-turbine generator, which was calculated as the costs of generating electricity plus the expenses of purchasing electricity minus the revenues from selling electricity. The mixed integer-linear programming problem was solved in two steps by applying GA and the interior point algorithm, respectively.

The procedure for implementing GA is illustrated in Fig. 2.2. GA is often applied to optimize the values of parameters in machine learning algorithms, or to solve

Fig. 2.2 Procedure for implementing GA

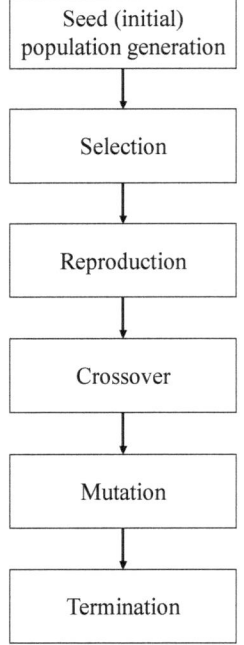

mathematical programming problems formulated for planning or controlling manufacturing systems. From this point of view, GA is a relatively indirect AI technology, so it is not necessarily an AI technology that lean manufacturing practitioners should learn first.

2.2.4 Artificial Neural Networks

ANNs are one of the most popular and widely used AI technologies. There are various types of ANNs, e.g., autoencoders, multilayer perceptrons, feedforward neural networks (or backpropagation networks), restricted Boltzmann machines, convolutional neural networks, spiking neural networks, long short-term memory networks, recurrent neural networks, self-organizing maps, etc. These ANNs support supervised, unsupervised, and reinforcement machine learning. ANNs are also combined with other AI technologies (such as GA, fuzzy logic, and deep learning).

Estimating the unit cost of a product is a critical task for manufacturers. However, improving the estimation accuracy is not easy because the unit cost of a product decreases with a learning process that involves considerable uncertainty. To overcome this difficulty, Chen et al. [33] proposed an agent-based fuzzy collaborative intelligence approach, in which agents autonomously fitted an uncertain unit cost learning process from different perspectives to estimate future unit costs. The estimation results by all agents were aggregated using fuzzy intersection (FI) [34]. If the entropy of the aggregation result was below a pre-specified threshold, a BPN was constructed to defuzzify the aggregation result to get a single representative value.

An ANN with multiple hidden layers is considered a deep neural network (DNN). A DNN may have the following advantages over an ANN [35]:

- The forecasting accuracy achieved using a DNN may be higher than that achieved using an ANN.
- A DNN may require a fewer number of nodes than an ANN to achieve the same forecasting accuracy.
- The training process of a DNN may be much more efficient than that of an ANN.

Wang et al. [36] constructed a two-dimensional long short-term memory (LSTM) network with multiple memory units to predict the cycle time of a wafer lot. The LSTM network was a deep recurrent neural network [37]. By deep learning and considering the correlation between network parameters, the estimation accuracy was improved.

Self-organizing maps (SOM) are commonly used to cluster and visualize high-dimensional data. A SOM projects high-dimensional data onto a two-dimension grid while roughly preserving the nonlinear dependencies between these dimensions. Alhoniemi et al. [38] constructed a SOM to analyze the correlation between the readings of eleven sensors of the continuous pulp digester of a pulp mill and product quality. In this way, they found a better way to set up the continuous pulp digester to improve product quality.

An adaptive network-based fuzzy inference system (ANFIS) is a combination of ANN and FIS, which provides an automatic method to enumerate all possible fuzzy inference rules from input variables. An example is illustrated in Fig. 2.3 with six input variables fuzzily partitioned into 3, 2, 3, 2, 2, and 2 linguistic terms, respectively. In theory, there will be at most 144 fuzzy inference rules. After training, only fuzzy inference rules that can effectively predict the output are left. ANFIS has been extensively applied in many fields. Fazlic et al. [39] constructed an ANFIS to predict the tar content of a cigarette product. Inputs to the ANFIS include the diameter, filter ventilation, nicotine, and carbon monoxide of the cigarette product, and each input variable was fuzzily partitioned into seven linguistic terms. However, the fuzzy partitioning results were not optimized. Therefore, a GA was also proposed by them to tune the membership functions of the linguistic terms to further improve the prediction performance.

The die yield of a wafer is the percentage of good dies on it. However, that can only be determined after packaging and final testing. By then, it's a little too late, and a lot of packaging and final test costs have already been spent. For this reason, estimating the die yield of a wafer based on the defect pattern on it (that is big data) is an important task, which is often accomplished using a convolutional neural network (CNN) [40]. A CNN has three types of layers: convolutional layers, pooling layers, and fully connected layers [41]. The first layer of a CNN is usually a convolutional

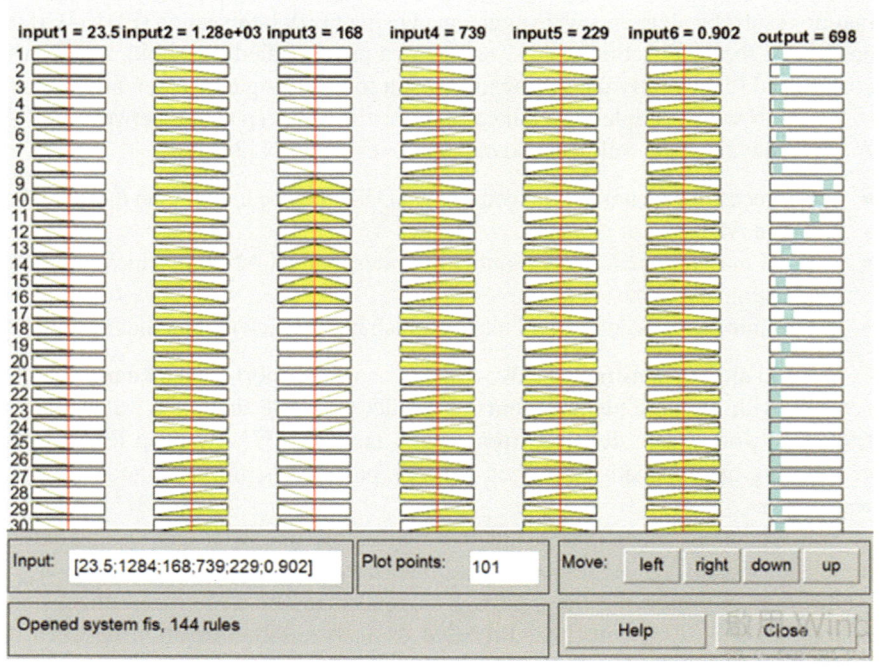

Fig. 2.3 Enumerating fuzzy inference rules

layer. It is connected to more convolutional layers or pooling layers. The last layer of the CNN is a fully connected layer. As the number of layers increases, the complexity of the CNN increases, allowing it to recognize more parts of the image. The first layer of the CNN focuses on simple features such as color and edges. As the image data passes through the CNN, other parts of the image, such as shapes, are recognized, and eventually the CNN is able to recognize the image.

2.2.5 Bayesian Approaches

Jones et al. [42] constructed a Bayesian network to model the effects of certain events on the failure risk of a manufacturing system. As new data or information became available, the probabilities of these events and their impacts on the failure risk of the manufacturing system were updated accordingly, allowing for more accurate failure risk predictions.

Estimating and reducing the variance in production results is critical to improving production efficiency and product quality. Existing methods for identifying the sources of variation often assume a sufficiently large amount of measurement data, which is not necessarily true. To address this issue, a Bayesian approach was proposed by Lee et al. [43]. First, the effects of possible sources of variation were modeled based on domain knowledge. These models were modified when more actual data or information became available. Then, the effects of these variation sources were re-evaluated.

In a wafer fab, various sensors are installed in machines to monitor their working status, resulting in a large amount of data, known as big data. For this reason, many big data analytics methods have been employed to analyze such data to generate valuable information for production planning and control. Yang and Lee [44] constructed a Bayesian belief network (BBN) to study the causal relationships between various conditions of a machine and evaluate their impact on wafer quality. These relationships were illustrated by the network structure, and the impact was quantified by conditional probabilities in the model.

2.2.6 Fuzzy Logic

Job scheduling is an important but difficult task for a wafer fab. To further improve the performance of job scheduling in a wafer fab, Chen [45] proposed a fuzzy-neural dynamic bottleneck detection (DBD) approach. The fuzzy-neural DBD approach is an improvement on the traditional DBD approach with some major changes. First, considering the uncertainty of job classification, fuzzy partition was applied to classify jobs into different categories. Second, the fuzzy c-means and fuzzy backpropagation network (FCM-FBPN) approach [46] was applied to estimate the remaining cycle time of a job. Third, the heuristics in the traditional DBD method were replaced with more advanced and flexible dispatching rules, such as the shortest cycle time until next bottleneck (SCNB) rule and the four-factor bi-criteria nonlinear fluctuation smoothing (4f-biNFS) rule [47].

Chen [48] established a ubiquitous manufacturing system of multiple 3D printing facilities, in which the printing time of a 3D object was defined as a fuzzy number to account for its uncertainty. Then, a fuzzy mixed integer-linear programming (FMILP) model and a fuzzy mixed integer-quadratic (FMIQP) model were optimized, respectively, to balance the loads on 3D printing facilities and plan the shortest delivery path.

After wafers are fabricated, some have to be scrapped because of too many defects. As a result, the energy used to fabricate these scrapped wafers is wasted. In order to minimize the waste of energy during wafer fabrication, Wang et al. [49] proposed a fuzzy nonlinear programming (FNLP) approach. They first modeled the process of resolving the quality problems of a product as a fuzzy yield learning process. Then, the energy saved by the yield learning process was quantified. Subsequently, an FNLP model was formulated and optimized to minimize the total energy consumption in the wafer fab by optimizing the product mix.

The application of specific AI technologies in manufacturing is affected by the following factors:

- Ease of understanding;
- Ease of communication;
- Ease of application;
- Effectiveness;
- Efficiency;
- Costs and expenses;
- Suitability for the application.

as illustrated in Fig. 2.4.

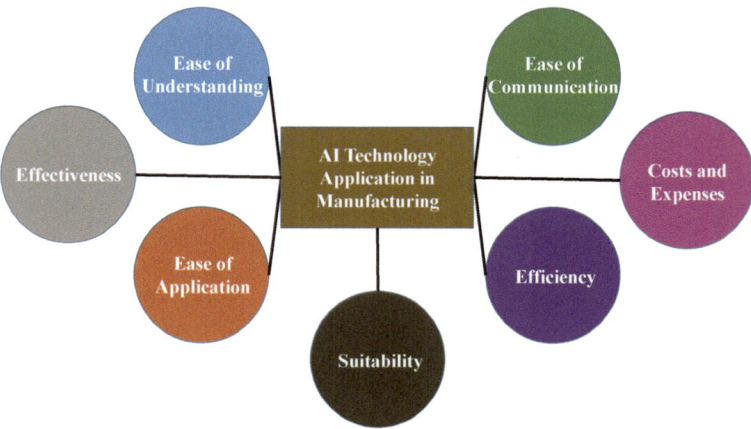

Fig. 2.4 Factors considered in choosing AI technology applications for a manufacturing system

2.2.7 3D Printing

Three-dimensional (3D) is to create a 3D object by forming successive layers of materials controlled by a computer-aided design and manufacturing (CAD/M) system. 3D printing was proposed by Nagoya Municipal Industrial Research Institute [50]. Acceptable 3D model file formats include.STL,.OBJ,.VRML,.PLY, and .ZIP. Chen and Lin [51] established a ubiquitous manufacturing system of multiple 3D printing facilities that distributed an order for 3D objects among multiple 3D printing facilities to minimize the makespan (i.e., longest cycle time).

3D printing has transformed from a tool for product research and development to a mechanism for mass production. Espera et al. [52] provides a detailed review of related reports and current practices on the application progress of 3D printing in electronics manufacturing by defining methods and protocols, reviewing various 3D printing methods, and describing the state-of-the-art in 3D printed electronics and its future.

4D printing technology is an evolution of 3D printing, in which the fourth dimension is time. In other words, using 4D printing technology, 3D printed products will change their shape or other properties as the environment (including temperature, brightness, humidity, etc.) changes [53].

2.2.8 Random Forests

A random forest is a classification algorithm consisting of many decision trees [54]. Each decision tree can be used to classify a new object. Then, the classification results by all decision trees are aggregated. Finally, the new object is classified into the group recommended by most decision trees. Like CARTs, random forests can be applied for prediction purposes.

Many wireless communication, sensing, and robotics devices are used simultaneously in smart factories. Therefore, smart factories are energy-intensive. To predict the power consumption of a smart factory, Sathishkumar et al. [55] applied four methods, including linear regression, radial kernel support vector machine, gradient boosting machine, and RF. Factors considered in predicting the power consumption of the smart factory included the lag and lead values of the factors, CO_2 emission, and load type. According to the experimental results, RF achieved the best prediction accuracy by minimizing root mean squared error.

For a battery manufacturing factory, Liu et al. [56] applied RF to analyze the effects of three intermediate product features from the mixing stage and one product parameter from the coating stage on the electrode active material mass load and porosity of a battery. According to the experimental results, the RF method not only reliably classified electrode properties, but also effectively quantified the effects of manufacturing features and their correlations.

2.3   AI Applications in Lean Manufacturing

2.3.1   Motivation

In the past, the application of AI was not the focus of lean manufacturing systems. However, the incoming AI boom, coupled with the increasing maturity of AI technologies, makes the application of AI extremely attractive for lean manufacturing systems [57].

Why does a lean manufacturing system need the application of AI technologies? On one hand, a lean manufacturing system can not only reduce wastes and inventory, but also respond to customer needs more immediately. On the other hand, AI technologies have been extensively applied to factories to optimize machine settings, solve production sequencing and scheduling problems, detect possible product quality problems, and diagnose the health of a machine. Applying AI to lean manufacturing is a feasible way to combine the advantages of these two disciplines.

In our view, lean manufacturing aims to operate a manufacturing system in a way that humans can recognize, while the applications of AI in manufacturing systems have developed to become more and more complicated, which may be beyond the scope of human understanding. The application of AI in lean manufacturing seems to be seeking the middle point between the two extremes or a practice to combine the strengths of both.

One concern is whether the application of AI technologies replaces the role of humans in the process of lean improvement. In the view of Devereaux [4], the application of AI technologies only provides more real-time information and better problem-solving tools for people in a factory.

2.3.2   Application Procedure

The following procedure can be followed in applying AI technologies to a lean manufacturing system (see Fig. 2.5):

Step 1. Form a team in a lean manufacturing system to conceive advanced data, control, and management, which may require the intervention of AI technologies.

Step 2. Team members list potentially useful AI technologies with the aid of an AI technology consultant.

Step 3. Team members assess these AI technologies. Factors considered in the assessment process are summarized in Fig. 2.6. The ease of understanding, communication, and application will be the key concern. In addition, lean manufacturing pursues continuous improvement [57]. Therefore, efficiency may not matter, since it is not necessary for an AI technology application to be effective immediately.

Step 4. Team members choose suitable AI technologies.

Step 5. Operators and engineers in the learn manufacturing system learn these AI technologies.

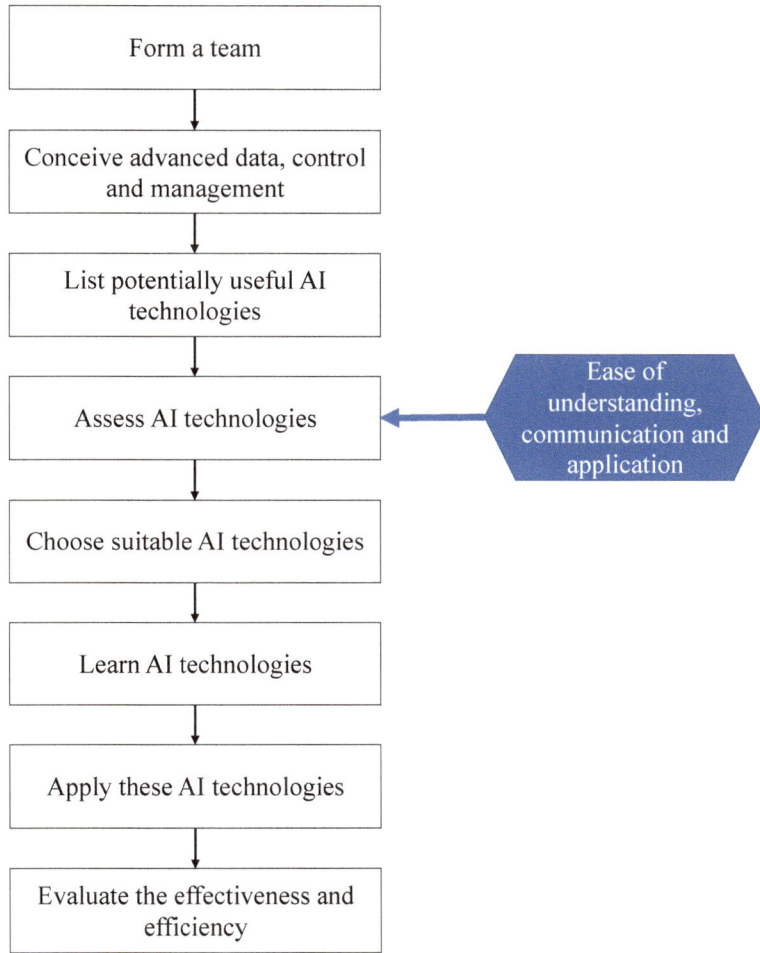

Fig. 2.5 Procedure for introducing AI technology applications into a lean manufacturing system

Step 6. Operators and engineers apply these AI technologies in the lean manufacturing system.

Step 7. Evaluate the effectiveness and efficiency of AI technology applications.

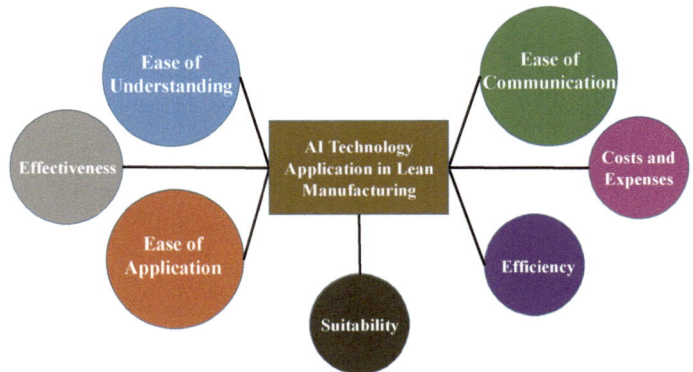

Fig. 2.6 Factors considered in assessing AI technology applications to lean manufacturing

2.3.3 Current Practice and Problems

Figure 2.7 provides statistics on the popularity of AI technology applications to lean manufacturing. Most commonly applied AI technologies in lean manufacturing include machine learning, inference, and fuzzy logic.

The application of AI technologies requires investment in information software and hardware as well as technical knowledge. These investments are usually not cheap. As far as the impact is concerned, the applications of AI technologies are often limited to large-scale manufacturing systems. However, according to Devereaux' view [4], AI technologies are potentially applicable in small manufacturing systems that pursue lean manufacturing. The reason is that a lean manufacturing system

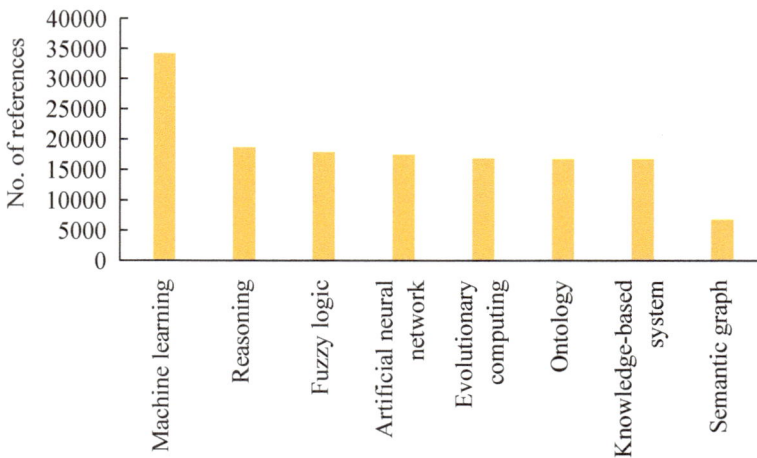

Fig. 2.7 Number of references about AI applications to lean manufacturing from 2010 to 2022 (*Data source* Google Scholar)

pursues continuous improvement. Once a problem is found, it is necessary to find out and eliminate possible causes, which leads to the interruption of operations. The application of AI technologies helps to detect, predict, and solve these problems without interrupting operations.

So far, there have been some literature or reports discussing the application of AI in lean manufacturing. However, current practice has the following problems:

- Some literature or reports did not specifically quantify the value of applying AI to lean manufacturing.
- Some literature or reports exaggerated the benefits brought by the application of AI.
- Most of the literature or reports present a lot of details of information technologies or systems, which are difficult for lean manufacturing practitioners to understand, communicate, and/or accept.

2.3.4 Examples

The application of advanced data analysis methods to find out directions that can reduce wastes is considered to be the starting point for the application of AI in a lean manufacturing system, because this approach does not require hardware investment and knowhow, can be applied quickly, and will not interfere with existing operations. Devereaux [4] suggested that a lean production system adopts machine learning as the beginning of AI applications. These applications are based on already-installed machine sensors, existing information software and hardware, and the expertise of data analysts.

Susilawati et al. [58] proposed a fuzzy many-criteria scoring approach to evaluate the leanness of a manufacturing system. The 66 criteria were divided into six categories: customer issues, supplier issues, manufacturing and internal business, research and development, learning perspectives, and investment priorities, as summarized in Table 2.1. AI technologies applicable to improving the performances in optimizing these criteria are also listed. However, too many evaluation criteria may dilute the excellence of a lean manufacturing system in a specific aspect.

In the view of Popa et al. [59], data mining can be applied to identify possible targets for improvement. Then, an expert system can be established to estimate the effects of improving these targets on the competitiveness of the manufacturing system.

Antosz et al. [60] collected the data of 150 lean enterprises, and then applied AI techniques (including CART, rough set theory, and GAs) to mine decision rules/trees to estimate the performance of a lean enterprise, in terms of the overall equipment effectiveness (OEE), based on its attributes.

Küfner et al. [61] constructed an ANN to classify the status of a machine. In this way, preventive maintenance can be carried out before the machine may be abnormal, so as to reduce the waste of production capacity caused by unexpected machine downs.

Table 2.1 Criteria for evaluating the leanness of a manufacturing system and applicable AI technologies

Category	Criteria	Applicable AI technologies
Customer issues	• Number of product variety • Product quality • Response time • Guarantee and warranty • Product • Product cost • Delivery performance • Customer requirement analysis • Product customization • Number of certified suppliers	• GAs • Inductive learning • Instance-based learning • ANNs • Bayesian approaches • Fuzzy logic
Supplier issues	• Communication and suggestions to suppliers • Involving suppliers in new product development • Keeping long-term partnerships with suppliers • Eliminate distance suppliers from the manufacturer location • Maintain quality of product sent by suppliers • Reducing time to supply product • Attempt to reduce number of suppliers of most important parts/materials • Total supply cost evaluation • Presentation of procedures written or documented in the company • Supplier performance evaluation	• GAs • Inductive learning • Instance-based learning • ANNs • Bayesian approaches • Fuzzy logic

(continued)

Table 2.1 (continued)

Category	Criteria	Applicable AI technologies
Manufacturing and internal business	• Stup time reduction • Work standardization • Cellular manufacturing • Error proofing • Value identification • Total productive maintenance • Shop floor organization • Total quality management • Cycle time reduction • Multifunctional work force • Work delegation • Employee evaluation • Bonus for the best employee performance • Manufacturing cycle time • Finished goods inventory • Raw material inventory • Equipment utilization • Labor utilization • Time spent on engineering change • Operation complexity • Defects in products • Excessive lead time • Excessive movement • Excessive scrap • Idleness of workers • Inappropriate processing • Machine down time • Non-utilization creativity of manpower • Poor fund management • Pull flow control • Production scheduling • Lot size reduction	• GAs • Inductive learning • Instance-based learning • ANNs • Bayesian approaches • Fuzzy logic

(continued)

Table 2.1 (continued)

Category	Criteria	Applicable AI technologies
Research and development	• Part standardization • Concurrent engineering • Design for manufacture • Lead time reduction of product • Development • Market research	• GAs • Inductive learning • Instance-based learning • ANNs • Bayesian approaches • Fuzzy logic
Learning perspectives	• Training for employees to do three or more jobs (multi-skilled) • Use of visual management or aids • Number of hours of training given to new employed personnel	NA
Investment priority	• Research and development • Automation processes • Training of employees • Market research • Procurement of new machinery • Procurement of advertising	• Inductive learning • Instance-based learning • Bayesian approaches • Fuzzy logic

With the increasing awareness of environmental protection, a factory that implements lean manufacturing must reduce the harm to the environment while reducing wastes in the factory. Therefore, when evaluating whether a factory is lean enough, environmental-related criteria must also be considered. Vahabi Nejat et al. [62] applied the fuzzy complex proportional assessment (fuzzy COPRAS) method to compare the lean levels of five Indian textile factories. The criteria considered in the comparison process were divided into six categories, including environmental protection and energy management. A fuzzy COPRAS method [63] considers the performances of maximizing and minimizing criteria in different ways. If there are alternatives that perform particularly well in optimizing minimizing criteria, the priorities of other alternatives will be significantly reduced.

Ante et al. [64] designed a key performance indicator (KPI) tree with the top levels showing the performances of a manufacturing system in various aspects, which were called key performance results. Below the top levels were KPIs organized from large items (at higher levels) to small items (at lower levels).

Pourjavad et al. [65] established a Mamdani FIS network to evaluate the performance of a lean manufacturing system, for which eleven criteria were considered. These criteria were grouped into five categories: productivity, output quantity, production costs, sales, and quality costs. For each category, a small Mamdani FIS was established to evaluate the overall performance of the category. Subsequently, the overall performances of all categories were input to a large Mamdani FIS to evaluate the overall performance of the lean manufacturing system. In addition to evaluating the performance of a lean manufacturing system, such a system can also measure the effect of an improvement (or kaizen) activity.

To implement SMED, Almomani et al. [66] proposed a multi-criteria decision-making method to select the best setting technique. The multi-criteria decision-making method was composed of analytical hierarchal process (AHP), preference selection index (PSI), and the technique for order preference by similarity to ideal solution (TOPSIS). First, AHP or PSI was applied to derive the priorities (or weights) of criteria. In AHP, the relative priorities of criteria were compared in pairs, while in PSI the priorities of criteria were proportional to the similarities of normalized performances [64]. Subsequently, the priorities of criteria derived using AHP and PSI were input to TOPSIS and WA, respectively, to evaluate the overall performances of alternatives. However, a decision-maker is often uncertain about the relative priority of one criterion over another. To model this uncertainty, AHP is usually replaced by fuzzy analytic hierarchy process (FAHP), where relative priorities are represented by fuzzy numbers. Subsequently, fuzzy geometric mean (FGM), fuzzy extent analysis (FEA), alpha-cut operations (ACO), or fuzzy relationships are applied to derive the priority of each criterion. Each of these methods has advantages and disadvantages [67].

2.4 Considerations in Introducing AI into Lean Manufacturing

The following recommendations can serve as a reference for lean manufacturing practitioners when considering the introduction of AI:

- Does a lean manufacturing system require the application of artificial intelligence? The answer may be no. With traditional lean manufacturing, additional resources may be wasted. Nonetheless, judging from past cases, the application of AI in lean manufacturing systems, rather than simply chasing popularity, is worth trying.
- AI technology applications often represent opportunities for improvement rather than solutions to existing problems.
- Lean manufacturing practitioners should learn the basic knowledge of AI technologies such as (ordinary) fuzzy logic, feedforward neural networks (FNNs), CBR, and other bottom-level AI technologies (see Fig. 2.8).
- Within each category of AI technologies, choose the simplest form.
- However, when the relevant software application is mature and easy to use, more difficult forms can be applied. For ease of use, a convenient user interface surpasses executable commands, which in turn surpasses program coding.
- The transparent management philosophy of lean manufacturing should be modified to accept black boxes in the manufacturing system (i.e., AI technology applications).

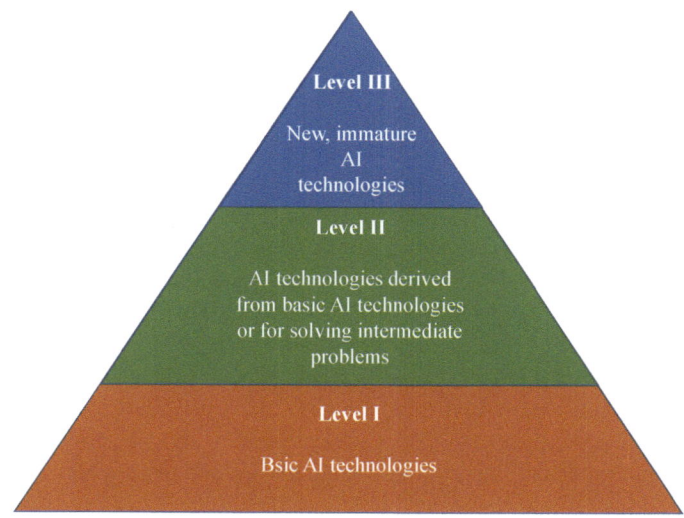

Fig. 2.8 Various levels of AI technologies

References

1. K.D. Pandl, S. Thiebes, M. Schmidt-Kraepelin, A. Sunyaev, On the convergence of artificial intelligence and distributed ledger technology: a scoping review and future research agenda. IEEE Access **8**, 57075–57095 (2020)
2. P. Perico, J. Mattioli, Empowering process and control in lean 4.0 with artificial intelligence, in *Third International Conference on Artificial Intelligence for Industries* (2020), pp. 6–9
3. C. Labreuche, S. Fossier, Explaining multi-criteria decision aiding models with an extended Shapley value, in *Proceedings of the Twenty-Seventh International Joint Conference on Artificial Intelligence* (2018), pp. 331–339
4. D. Devereaux, Smaller manufacturers get lean with artificial intelligence (2019). http://www.nist.gov/blogs/manufacturing-innovation-blog/smaller-manufacturers-get-leanartificial-intelligence
5. Y. Sun, L. Li, H. Shi, D. Chong, The transformation and upgrade of China's manufacturing industry in Industry 4.0 era. Syst. Res. Behav. Sci. **37**(4), 734–740 (2020)
6. P. Palensky, D. Bruckner, A. Tmej, T. Deutsch, Paradox in AI–AI 2.0: the way to machine consciousness, in *International Conference on IT Revolutions* (2008), pp. 194–215
7. Y.H. Pan, Heading toward artificial intelligence 2.0. Engineering **2**(4), 409–413 (2016)
8. P.J. Lisboa, AI 2.0: Augmented intelligence, data science and knowledge engineering for sensing decision support, in *Proceedings of the 13th International FLINS Conference* (2018), pp. 10–11
9. B.H. Li, B.C. Hou, W.T. Yu, X.B. Lu, C.W. Yang, Applications of artificial intelligence in intelligent manufacturing: a review. Front. Inform. Technol. Electron. Eng. **18**(1), 86–96 (2017)
10. A. Manghani, A primer on machine learning (2017). https://ce.uci.edu/pdfs/certificates/machine_learning_article.pdf
11. IBM, Supervised learning (2022). https://www.ibm.com/cloud/learn/supervised-learning
12. JavaTpoint, Unsupervised machine learning (2022). https://www.javatpoint.com/unsupervised-machine-learning
13. B. Dickson, What is semi-supervised machine learning? (2021). https://bdtechtalks.com/2021/01/04/semi-supervised-machine-learning/
14. B. Osiński, K. Budek, What is reinforcement learning? The complete guide (2018). https://deepsense.ai/what-is-reinforcement-learning-the-complete-guide/
15. T. Wuest, D. Weimer, C. Irgens, K.D. Thoben, Machine learning in manufacturing: advantages, challenges, and applications. Prod. Manuf. Res. **4**(1), 23–45 (2016)
16. L. Haldurai, T. Madhubala, R. Rajalakshmi, A study on genetic algorithm and its applications. Int. J. Comput. Sci. Eng. **4**(10), 139 (2016)
17. D. Graupe, *Principles of Artificial Neural Networks*, vol. 7 (World Scientific, 2013)
18. J. Mockus, *Bayesian Approach to Global Optimization: Theory and Applications*, vol. 37 (Springer Science & Business Media, 2012)
19. H.C. Wu, T. Chen, CART–BPN approach for estimating cycle time in wafer fabrication. J. Ambient. Intell. Humaniz. Comput. **6**(1), 57–67 (2015)
20. C. Wang, X.P. Tan, S.B. Tor, C.S. Lim, Machine learning in additive manufacturing: State-of-the-art and perspectives. Addit. Manuf. **36**, 101538 (2020)
21. S.C.H. Lu, D. Ramaswamy, P.R. Kumar, Efficient scheduling policies to reduce mean and variation of cycle time in semiconductor manufacturing plant. IEEE Trans. Semicond. Manuf. **7**(3), 374–388 (1994)
22. T.C. Chen, Y.C. Wang, Y.C. Lin, A fuzzy-neural system for scheduling a wafer fabrication factory. Int. J. Innov. Comput. Inform. Control **6**(2), 687–700 (2010)
23. A. Amindoust, S. Ahmed, A. Saghafinia, A. Bahreininejad, Sustainable supplier selection: a ranking model based on fuzzy inference system. Appl. Soft Comput. **12**(6), 1668–1677 (2012)
24. T. Madhusudan, J.L. Zhao, B. Marshall, A case-based reasoning framework for workflow model management. Data Knowl. Eng. **50**(1), 87–115 (2004)
25. A. González-Briones, J. Prieto, F. De La Prieta, E. Herrera-Viedma, J.M. Corchado, Energy optimization using a case-based reasoning strategy. Sensors **18**(3), 865 (2018)

26. J. Lim, M.J. Chae, Y. Yang, I.B. Park, J. Lee, J. Park, Fast scheduling of semiconductor manufacturing facilities using case-based reasoning. IEEE Trans. Semicond. Manuf. **29**(1), 22–32 (2015)
27. P.C. Chang, J.C. Hsieh, T.W. Liao, A case-based reasoning approach for due-date assignment in a wafer fabrication factory, in *International Conference on Case-Based Reasoning* (2001), pp. 648–659
28. S. Shigeo, A.P. Dillon. *A Revolution in Manufacturing: The SMED System* (Routledge, 2019)
29. R.J. Kuo, L.M. Lin, Application of a hybrid of genetic algorithm and particle swarm optimization algorithm for order clustering. Decis. Support Syst. **49**(4), 451–462 (2010)
30. T. Chen, C.W. Lin, Smart and automation technologies for ensuring the long-term operation of a factory amid the COVID-19 pandemic: an evolving fuzzy assessment approach. Int. J. Adv. Manuf. Technol. **111**(11), 3545–3558 (2020)
31. H. Kurniawan, T.D. Sofianti, A.T. Pratama, P.I. Tanaya, Optimizing production scheduling using genetic algorithm in textile factory. J. Syst. Manage. Sci. **4**(4), 27–44 (2014)
32. Y.Y. Hong, P.S. Yo, Novel genetic algorithm-based energy management in a factory power system considering uncertain photovoltaic energies. Appl. Sci. **7**(5), 438 (2017)
33. T. Chen, Estimating unit cost using agent-based fuzzy collaborative intelligence approach with entropy-consensus. Appl. Soft Comput. **73**, 884–897 (2018)
34. T. Chen, Y.C. Lin, A fuzzy-neural system incorporating unequally important expert opinions for semiconductor yield forecasting. Internat. J. Uncertain. Fuzziness Knowl.-Based Syst. **16**(01), 35–58 (2008)
35. T.C.T. Chen, Y.C. Wang, Fuzzy dynamic-prioritization agent-based system for forecasting job cycle time in a wafer fabrication plant. Complex Intell. Syst. **7**(4), 2141–2154 (2021)
36. J. Wang, J. Zhang, X. Wang, Bilateral LSTM: A two-dimensional long short-term memory model with multiply memory units for short-term cycle time forecasting in re-entrant manufacturing systems. IEEE Trans. Indus. Inform. **14**(2), 748–758 (2017)
37. G. Montavon, W. Samek, K.R. Müller, Methods for interpreting and understanding deep neural networks. Digit Signal Process **73**, 1–15 (2018)
38. E. Alhoniemi, J. Hollmén, O. Simula, J. Vesanto, Process monitoring and modeling using the self-organizing map. Integr. Comput. Aided Eng. **6**(1), 3–14 (1999)
39. L.B. Fazlic, Z. Avdagic, I. Besic, GA-ANFIS expert system prototype for detection of tar content in the manufacturing process, in *2015 38th International Convention on Information and Communication Technology, Electronics and Microelectronics* (2015), pp. 1194–1199
40. J. Moyne, J. Samantaray, M. Armacost, Big data capabilities applied to semiconductor manufacturing advanced process control. IEEE Trans. Semicond. Manuf. **29**(4), 283–291 (2016)
41. IBM Cloud Education, Convolutional neural networks (2020). https://www.ibm.com/cloud/learn/convolutional-neural-networks
42. B. Jones, I. Jenkinson, Z. Yang, J. Wang, The use of Bayesian network modelling for maintenance planning in a manufacturing industry. Reliab. Eng. Syst. Saf. **95**(3), 267–277 (2010)
43. J. Lee, J. Son, S. Zhou, Y. Chen, Variation source identification in manufacturing processes using Bayesian approach with sparse variance components prior. IEEE Trans. Autom. Sci. Eng. **17**(3), 1469–1485 (2020)
44. L. Yang, J. Lee, Bayesian Belief Network-based approach for diagnostics and prognostics of semiconductor manufacturing systems. Robot. Comput.-Integr. Manuf. **28**(1), 66–74 (2012)
45. T. Chen, A fuzzy-neural DBD approach for job scheduling in a wafer fabrication factory. Int. J. Innov. Comput. Inform. Control **8**(6), 4024–4044 (2012)
46. T. Chen, Y.C. Wang, H.C. Wu, A fuzzy-neural approach for remaining cycle time estimation in a semiconductor manufacturing factory—a simulation study. Int. J. Innov. Comput. Inform. Control **5**(8), 2125–2139 (2009)
47. T. Chen, Y.C. Wang, Y.C. Lin, A bi-criteria four-factor fluctuation smoothing rule for scheduling jobs in a wafer fabrication factory. Int. J. Innov. Comput. Inform. Control **6**(10), 4289–4304 (2009)

48. T.C.T. Chen, Fuzzy approach for production planning by using a three-dimensional printing-based ubiquitous manufacturing system. AI EDAM **33**(4), 458–468 (2019)
49. Y.C. Wang, M.C. Chiu, T. Chen, A fuzzy nonlinear programming approach for planning energy-efficient wafer fabrication factories. Appl. Soft Comput. **95**, 106506 (2020)
50. H. Kodama, A scheme for three-dimensional display by automatic fabrication of three-dimensional model. IEICE Trans. Electron. J. **64**-C(4), 237–241 (1981)
51. T.C.T. Chen, Y.C. Lin, A three-dimensional-printing-based agile and ubiquitous additive manufacturing system. Robot. Comput.-Integr. Manuf. **55**, 88–95 (2019)
52. A.H. Espera, J.R.C. Dizon, Q. Chen, R.C. Advincula, 3D-printing and advanced manufacturing for electronics. Progress Additive Manuf. **4**(3), 245–267 (2019)
53. Q. Ge, A.H. Sakhaei, H. Lee, C.K. Dunn, N.X. Fang, M.L. Dunn, Multimaterial 4D printing with tailorable shape memory polymers. Sci. Rep. **6**(1), 1–11 (2016)
54. T. Yiu, Understanding random forest (2019). https://towardsdatascience.com/understanding-random-forest-58381e0602d2
55. V.E. Sathishkumar, M. Lee, J. Lim, Y. Kim, C. Shin, J. Park, Y. Cho, An energy consumption prediction model for smart factory using data mining algorithms. KIPS Trans. Softw. Data Eng. **9**(5), 153–160 (2020)
56. K. Liu, X. Hu, H. Zhou, L. Tong, W.D. Widanage, J. Marco, Feature analyses and modeling of lithium-ion battery manufacturing based on random forest classification. IEEE/ASME Trans. Mechatron. **26**(6), 2944–2955 (2021)
57. M.L. George Sr, D.K. Blackwell, D. Rajan, *Lean Six Sigma in the Age of Artificial Intelligence: Harnessing the Power of the Fourth Industrial Revolution* (McGraw-Hill Education, 2019)
58. A. Susilawati, J. Tan, D. Bell, M. Sarwar, Fuzzy logic based method to measure degree of lean activity in manufacturing industry. J. Manuf. Syst. **34**, 1–11 (2015)
59. A. Popa, R. Ramos, A.B. Cover, C.G. Popa, Integration of artificial intelligence and lean sigma for large field production optimization: Application to Kern River Field, in *SPE Annual Technical Conference and Exhibition* (2005)
60. K. Antosz, L. Pasko, A. Gola, The use of artificial intelligence methods to assess the effectiveness of lean maintenance concept implementation in manufacturing enterprises. Appl. Sci. **10**(21), 7922 (2020)
61. T. Küfner, T.H.J. Uhlemann, B. Ziegler, Lean data in manufacturing systems: Using artificial intelligence for decentralized data reduction and information extraction. Procedia CIRP **72**, 219–224 (2018)
62. S. Vahabi Nejat, S. Avakh Darestani, M. Omidvari, M.A. Adibi, Evaluation of green lean production in textile industry: a hybrid fuzzy decision-making framework. Environ. Sci. Pollut. Res. **29**(8), 11590–11611 (2022)
63. A. Alinezhad, J. Khalili, COPRAS method. Internat. Ser. Oper. Res. Manage. Sci. **277**, 87–91 (2019)
64. G. Ante, F. Facchini, G. Mossa, S. Digiesi, Developing a key performance indicators tree for lean and smart production systems. IFAC-PapersOnLine **51**(11), 13–18 (2018)
65. E. Pourjavad, R.V. Mayorga, A comparative study and measuring performance of manufacturing systems with Mamdani fuzzy inference system. J. Intell. Manuf. **30**(3), 1085–1097 (2019)
66. M.A. Almomani, M. Aladeemy, A. Abdelhadi, A. Mumani, A proposed approach for setup time reduction through integrating conventional SMED method with multiple criteria decision-making techniques. Comput. Ind. Eng. **66**(2), 461–469 (2013)
67. K. Maniya, M.G. Bhatt, A selection of material using a novel type decision-making method: preference selection index method. Mater. Des. **31**(4), 1785–1789 (2010)

Chapter 3
AI Applications to Kaizen Management

3.1 Kaizen Activities in Lean Manufacturing

Kaizen (or improvement) activities are the core of lean manufacturing [1]. The following related topics are discussed in this chapter:

- **Leanness**, which is the outcome of all kaizen activities.
- **5S**, including kaizen activities for improving the working environment;
- **Predictive maintenance**, including kaizen activities for improving the reliability (or availability) of equipment.
- **Cycle time reduction**, which is a major task when improving the value stream map (VSM) of a manufacturing system.

AI technologies that are applicable to these topics are introduced.

3.2 Leanness of a Manufacturing System

The leanness of a manufacturing system is the outcome of all kaizen activities in the manufacturing system. A manufacturing system is completely lean if there is no waste in it. However, that is not easy to know. For this reason, there have been a number of methods proposed in the literature to evaluate the leanness of a manufacturing system [2]. The common idea of these methods is that a manufacturing system should be highly lean if it performs well on many kaizen activities.

For example, Vimal and Vinodh [3] applied **fuzzy weighted average (FWA)** to evaluate the leanness of a manufacturing system, in which 28 criteria were considered. The performances of the manufacturing system in optimizing these criteria were modeled as fuzzy numbers, such as triangular fuzzy numbers (TFNs) [4]. A TFN $\tilde{A} = (A_1, A_2, A_3)$ is illustrated in Fig. 3.1, where A_1 is the most likely time; A_2 and A_3 represent the shortest and longest times, respectively. Such notation is easy

© The Author(s), under exclusive license to Springer Nature Switzerland AG 2022
T.-C. T. Chen and Y.-C. Wang, *Artificial Intelligence and Lean Manufacturing*,
SpringerBriefs in Applied Sciences and Technology,
https://doi.org/10.1007/978-3-031-04583-7_3

Fig. 3.1 TFN

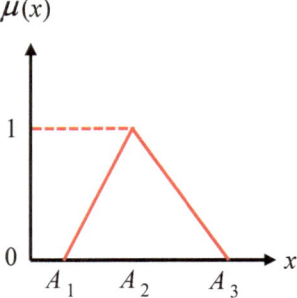

to understand and communicate. The membership function of $\tilde{A}$ is given by

$$\mu_{\tilde{A}}(x) = \max\left(\min\left(\frac{x - A_1}{A_2 - A_1}, \ \frac{x - A_3}{A_2 - A_3}\right), \ 0\right) \tag{3.1}$$

Some arithmetic operations on TFNs are summarized below [5].

- Fuzzy addition:

$$(A_1, \ A_2, \ A_3)(+)(B_1, \ B_2, \ B_3) = (A_1 + B_1, \ A_2 + B_2, \ A_3 + B_3) \tag{3.2}$$

- Fuzzy subtraction:

$$(A_1, \ A_2, \ A_3)(-)(B_1, \ B_2, \ B_3) = (A_1 - B_3, \ A_2 - B_2, \ A_3 - B_1) \tag{3.3}$$

- Scalar multiplication:

$$k(A_1, \ A_2, \ A_3) = (kA_1, \ kA_2, \ kA_3) \quad \text{if } k \geq 0 \tag{3.4}$$

- Fuzzy multiplication:

$$(A_1, \ A_2, \ A_3)(\times)(B_1, \ B_2, \ B_3) \cong (A_1 B_1, \ A_2 B_2, \ A_3 B_3) \quad \text{if } A_1, \ B_1 \geq 0 \tag{3.5}$$

- Fuzzy division:

$$(A_1, \ A_2, \ A_3)(/)(B_1, \ B_2, \ B_3) \cong (A_1/B_3, \ A_2/B_2, \ A_3/B_1) \quad \text{if } A_1 \geq 0, \ B_1 > 0 \tag{3.6}$$

- Fuzzy maximum:

$$\text{Max}((A_1, \ A_2, \ A_3), \ (B_1, \ B_2, \ B_3)) = (\text{Max}(A_1, \ B_1), \ \text{Max}(A_2, \ B_2), \ \text{Max}(A_3, \ B_3)) \tag{3.7}$$

- Fuzzy minimum:

$$Min((A_1, A_2, A_3), (B_1, B_2, B_3)) = (Min(A_1, B_1), Min(A_2, B_2), Min(A_3, B_3)) \quad (3.8)$$

FWA evaluates the leanness of a manufacturing system in the following manner:

$$FWA(\text{Leanness}) = \frac{\sum_{i=1}^{n} (\tilde{w}_i(\times)\tilde{p}_i)}{\sum_{i=1}^{n} \tilde{w}_i} \quad (3.9)$$

where $\tilde{p}_i$ is the performance of the manufacturing system in optimizing criterion i; $\tilde{w}_i$ is the weight (or priority) of this criterion. If $\sum_{i=1}^{n} \tilde{w}_i = 1$ or the leanness of several manufacturing systems are to be compared, Eq. (3.9) can be simplified as

$$FWA(\text{Leanness}) = \sum_{i=1}^{n} (\tilde{w}_i(\times)\tilde{p}_i) \quad (3.10)$$

Example 3.1 The performances of a manufacturing system in optimizing several criteria [3] (illustrated in Fig. 3.2) are summarized in Table 3.1. The weights of these criteria are also provided in this table. Then, the leanness of the manufacturing system can be evaluated as

Fig. 3.2 Leanness of a manufacturing system

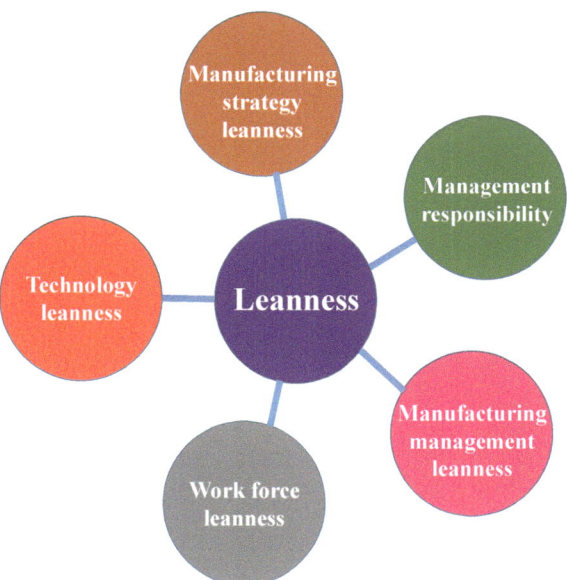

Table 3.1 Performances of a manufacturing system in optimizing several criteria

Criterion	Performance	Weight
Management responsibility	$(3.5, 4, 4.5)$	$(0.2, 0.25, 0.3)$
Manufacturing management leanness	$(2.8, 3.2, 3.7)$	$(0.1, 0.15, 0.2)$
Work force leanness	$(4.1, 4.5, 4.7)$	$(0.35, 0.4, 0.45)$
Technology leanness	$(3.3, 3.6, 4.0)$	$(0.1, 0.15, 0.2)$
Manufacturing strategy leanness	$(2.4, 2.7, 3.1)$	$(0, 0.05, 0.1)$

FWA(Leanness)

$= (3.5, \ 4, \ 4.5)(\times)(0.2, \ 0.25, \ 0.3)(+)(2.8, \ 3.2, \ 3.7)(\times)(0.1, \ 0.15, \ 0.2)$
$(+)(4.1, \ 4.5, \ 4.7)(\times)(0.35, \ 0.4, \ 0.45)(+)(3.3, \ 3.6, \ 4.0)(\times)(0.1, \ 0.15, \ 0.2)$
$(+)(2.4, \ 2.7, \ 3.1)(\times)(0, \ 0.05, \ 0.1)$
$= (0.7, \ 1, \ 1.35)(+)(0.28, \ 0.48, \ 0.74)(+)(1.435, \ 1.8, \ 2.115)(+)(0.33, \ 0.54, \ 0.8)$
$(+)(0, \ 0.135, \ 0.31)$
$= (2.745, \ 3.955, \ 5.315)$

The manufacturing system should compare the evaluated leanness with those of other manufacturing systems, or monitor whether it continues to improve.

The evaluation result can be converted into a single (i.e., crisp) value using defuzzification. To this end, the center-of-gravity (COG) method is widely applied [6]:

$$COG(\tilde{A}) = \frac{\int\limits_{x} x\mu_{\tilde{A}}(x)dx}{\int\limits_{x} \mu_{\tilde{A}}(x)dx} \tag{3.11}$$

The following theorem facilitates the calculation of Eq. (3.11).

Theorem 3.1 *The COG value of a TFN* $\tilde{A} = (A_1, \ A_2, \ A_3)$ *can be calculated as* [6]

$$COG(\tilde{A}) = \frac{A_1 + A_2 + A_3}{3} \tag{3.12}$$

In the previous example, the COG of the evaluated leanness is calculated as

$$COG(FWA(\text{Leanness}))$$
$$= \frac{2.745 + \ 3.955 + \ 5.315}{3}$$
$$= 4.005$$

Other types of fuzzy numbers, such as trapezoidal fuzzy numbers, generalized bell fuzzy numbers, Gaussian fuzzy numbers are also applicable.

3.3 AI Applications to 5S

5S emphasize making people do it [7]. In contrast, AI uses computers to perform tasks on behalf of people. Therefore, whether there is room for AI to be applied to 5S is an interesting issue.

Randhawa and Ahuja [8] constructed a **fuzzy inference system (FIS)** to evaluate the success of 5S implementation. The input variables of this FIS were the performances of conducting a series of 5S activities, and the output was the success degree of 5S implementation.

An FIS is composed of multiple fuzzy inference rules. An example is given as follows:

Rule 1:   If "tools are returned to fixed positions" is "very frequently" And "shop floor is thoroughly cleaned" is "very frequently" Then "success of 5S implementation" is "high".

This fuzzy inference rule has two premises corresponding to the two variables "tools are returned to fixed positions" and "shop floor is thoroughly cleaned", and one consequence "success of 5S implementation". All of them have linguistic values, such as {"very seldom", "seldom", "moderately", "frequently", "very frequently"} and {"very low", "low", "medium", "high", "very high"}. Fuzzy inference rules are easy to understand and communicate, making them suitable for lean manufacturing systems. The establishment of fuzzy inference rules may be based on experts' subjective experiences or from mining historical data.

The procedure for constructing an FIS is illustrated in Fig. 3.3.

Linguistic values can be mapped to TFNs by fuzzily partitioning the range of each variable into a series of TFNs that overlap with each other, as illustrated in Fig. 3.4.

The fuzzy inference rule is illustrated with the corresponding TFNs in Fig. 3.5.

Example 3.2 The production manager of a lean manufacturing system scored the performances of conducting 5S activities using values within [0, 5] as

"Tools are returned to fixed positions" = 4.2
"Shop floor is thoroughly cleaned" = 4.7

which are compared with the fuzzy inference rule in Fig. 3.5. The memberships of scores in the TFNs are derived as

$$\mu_{\text{very frequently}}(4.2) = \frac{4.2 - 4}{5 - 4} = 0.2$$

$$\mu_{\text{very frequently}}(4.7) = \frac{4.7 - 4}{5 - 4} = 0.7$$

Fig. 3.3 Procedure for
constructing an FIS

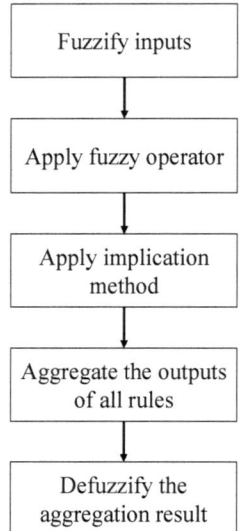

An "and" operator is used to combine the two premises. Therefore, the minimum
of their memberships is calculated as $\min(0.2, 0.7) = 0.2$, which is used to truncate
the consequence, as illustrated in Fig. 3.6. As a result, the evaluated success of 5S
implementation is a trapezoidal fuzzy number (TrFN) [9]. An FIS that operates in
this way is called a Mamdani FIS [10].

The estimation result using the fuzzy inference rule can be converted into a single
(i.e., crisp) value using defuzzification. To this end, the following theorem is helpful.

Theorem 3.2 *The COG value of the TrFN shown in* Fig. 3.7 *can be calculated as*
[11]

$$COG(\tilde{A}) = \frac{1}{3}\left(A_1 + A_2 + A_3 + A_4 - \frac{A_3A_4 - A_1A_2}{(A_3 + A_4) - (A_1 + A_2)}\right) \qquad (3.13)$$

In the previous example, the COG of the evaluated success of 5S implementation
can be derived as

$$COG(\text{Evaluated success of 5S implementation})$$
$$= \frac{1}{3}\left(6 + 6.3 + 8.7 + 9 - \frac{8.7 \cdot 9 - 6 \cdot 6.3}{(8.7 + 9) - (6 + 6.3)}\right)$$
$$= 7.5$$

If multiple fuzzy inference rules are applied to the collected data, their results are
aggregated using fuzzy union (FU) before defuzzification in a Mamdani FIS system
[12], as illustrated in Fig. 3.8:

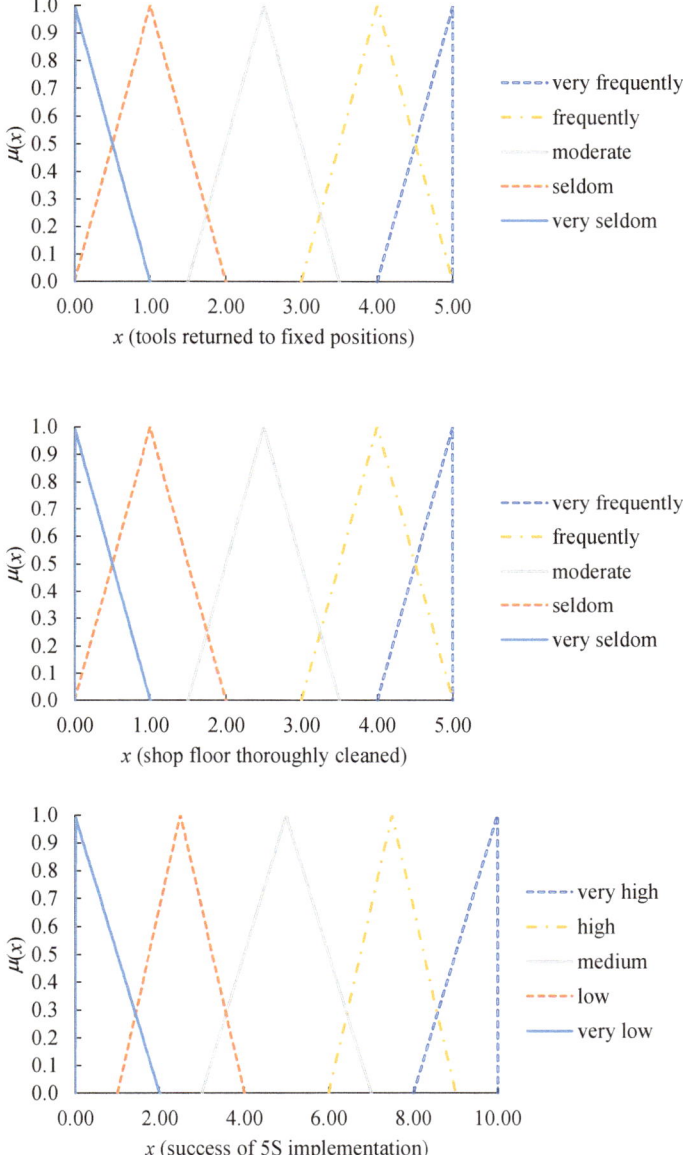

Fig. 3.4 Fuzzy partitioning results

$$\mu_{FU(\tilde{A}_i)}(x) = \max_i(\mu_{\tilde{A}_i}(x)) \qquad (3.14)$$

The aggregation result is a polygonal fuzzy number [13] which is not easy to defuzzify. To tackle this difficulty, the following theorems are helpful.

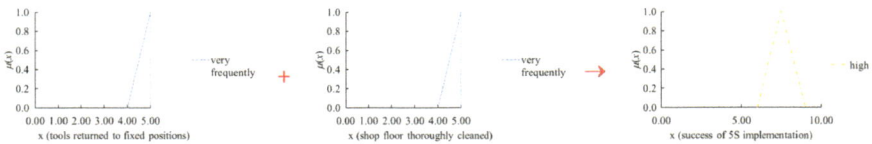

Fig. 3.5 A fuzzy inference rule

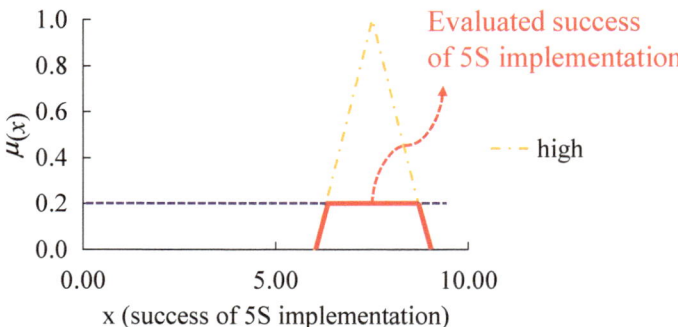

Fig. 3.6 Result of applying the fuzzy inference rule

Fig. 3.7 A TrFN

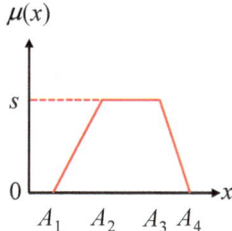

Theorem 3.3 *The integral of the TrFN shown in* Fig. 3.9 *can be calculated as* [14]

$$\int_{x_1}^{x_2} \mu_{\tilde{A}(x)}(x)dx = \frac{\mu_2 x_2^2 + \mu_1 x_2^2 - 2\mu_2 x_1 x_2 + \mu_1 x_1^2 - 2\mu_1 x_1 x_2 + \mu_2 x_1^2}{2(x_2 - x_1)} \quad (3.15)$$

Theorem 3.4 [14]

$$\int_{x_1}^{x_2} x\mu_{\tilde{A}(x)}(x)dx = \frac{2\mu_2 x_2^3 + \mu_1 x_2^3 - 3\mu_2 x_1 x_2^2 + \mu_2 x_1^3 + 2\mu_1 x_1^3 - 3\mu_1 x_1^2 x_2}{6(x_2 - x_1)} \quad (3.16)$$

The aggregation result can be divided into several parts, as illustrated in Fig. 3.10. Each part is similar to the TrFN in Fig. 3.9. Then, Theorems 3.3 and 3.4 can be applied to calculate the COG of the aggregation result.

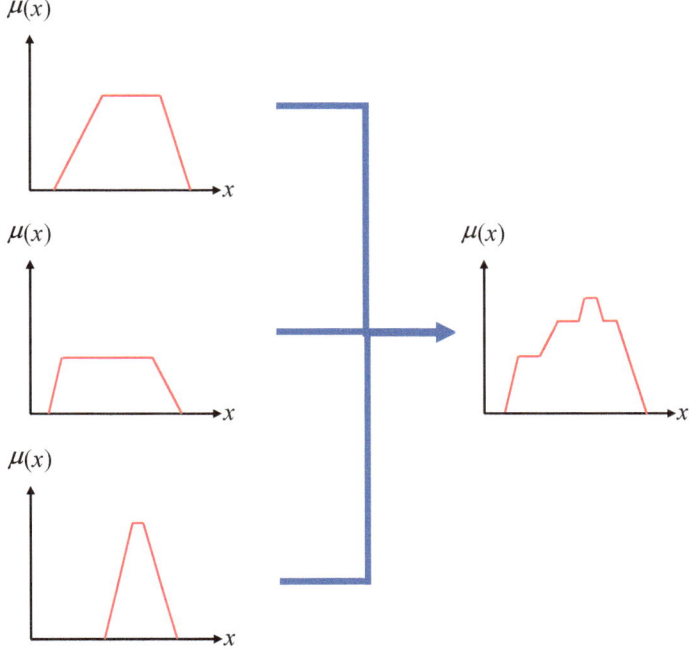

Fig. 3.8 Aggregating the results of multiple fuzzy inference rules

Fig. 3.9 A non-normal
TrFN

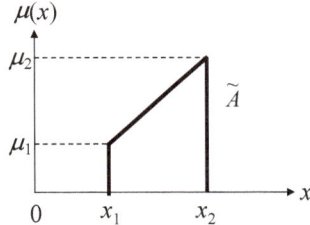

Fig. 3.10 Dividing the
aggregation results into
TrFNs

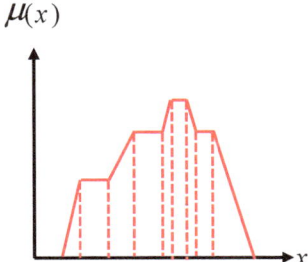

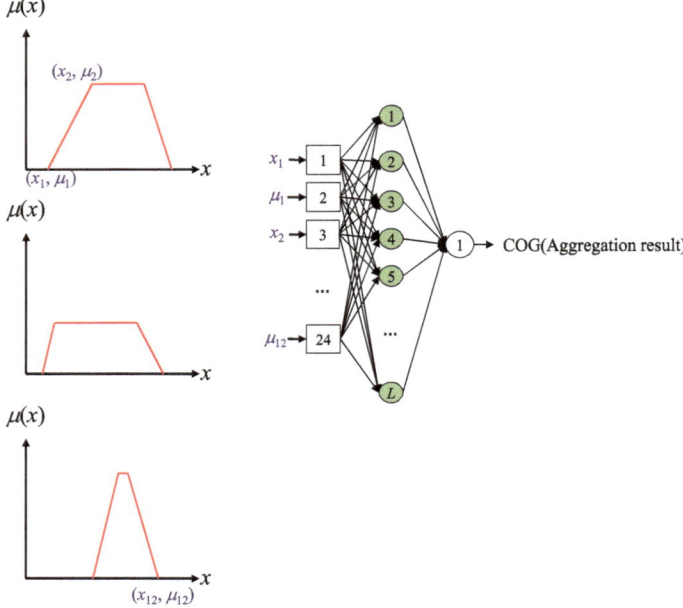

Fig. 3.11 ANN for evaluating the success of 5S implementation

To further facilitate the calculation, an artificial neural network (ANN) [3] can be constructed to estimate the COG of the aggregation result, so as to evaluate the success of 5S implementation. Inputs to the ANN are the results (i.e., the values and memberships of their endpoints) of all fuzzy inference rules, and the output is the estimated COG of the aggregation result (i.e., the evaluated success of 5S implementation), as illustrated in Fig. 3.11.

Existing FISs can be classified into three categories:

- Tsukamoto FISs: In a Tsukamoto FIS, the membership of the consequence in a fuzzy inference rule is set to the satisfaction level of premises. In this way, the output is derived. For example, in Example 3.2, the output from the fuzzy inference rule will be 6.3 (a pessimistic value) or 8.7 (an optimistic value) if the FIS is of Tsukamoto type, as shown in Fig. 3.12.

Subsequently, the weighted sum of the outputs of all rules is calculated as the final result, for which the weight of a rule is equal to its satisfaction level.

- Mamdani FISs: The FIS introduced above is a Mamdani FIS.
- Sugeno FISs: In a Sugeno FIS, the output from a fuzzy inference rule is a linear function of inputs. For example, in Example 3.2, the output from the fuzzy inference rule is

  1.4 · "tools are returned to fixed positions" + 0.6 · "shop floor is thoroughly cleaned"

Fig. 3.12 Output of the
fuzzy inference rule
(Tsukamoto FIS)

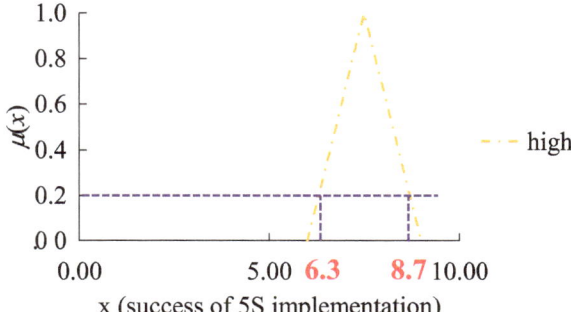

$$= 1.4 \cdot 4.2 + 0.6 \cdot 4.7$$
$$= 8.7$$

Subsequently, the weighted sum of the outputs of all rules is calculated as the
final result, for which the weight of a rule is equal to its satisfaction level.

3.4   AI Applications to Predictive Maintenance

Predictive maintenance is undoubtedly a critical improvement (or kaizen) activity in
lean manufacturing. **Predictive maintenance** typically employs sensors to monitor
the conditions of a machine in various aspects [15]. The relationship between these
conditions and the time until the next machine down is then fitted. The fitting result
may not be a single function, but rather represented as knowledge. To this end,
knowledge representation and reasoning (KRR) is helpful. KRR is a field of
AI that represents information about the world in a form that a computer system
can learn to solve complex tasks [16]. Based on such knowledge, measures can
be formulated to eliminate the causes of machine early failures, thereby increasing
machine availability and reliability.

An **FIS** can also be built to organize the knowledge generated from a predictive
maintenance application [3]. Such an FIS is composed of fuzzy inference rules like

Rule 1:   If "machine temperature" is "very high" and "noise level" is "high" Then
"time to the next failure" is "very short".

Rule 2:   If "machine temperature" is "low" or "noise level" is "low" then "time to
the next failure" is "medium".

Rule 3:   If "noise level" is "very high" then "time to the next failure" is "short".

When the "or" operator is used to combine the premises, the maximum of their
memberships is calculated [17]. In addition to predicting the time to the next failure
of a machine, an FIS system can also measure the effect of an improvement (or
kaizen) activity on postponing the next failure.

Example 3.3 A Mamdani FIS [10] is established to predict the next failure of a machine based on the monitoring results of temperature and noise sensors. First, the variables in the FIS are fuzzily partitioned. The partitioning results are shown in Fig. 3.13.

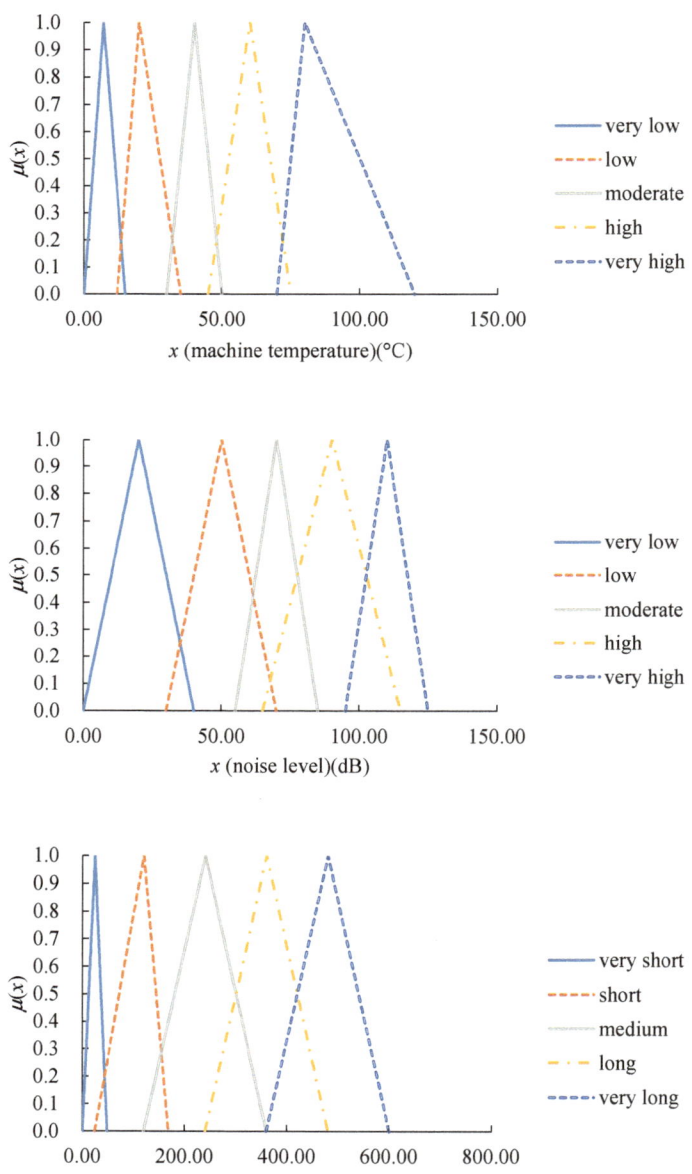

Fig. 3.13 Partitioning results of fuzzy variables

The fuzzy logic toolbox of MATLAB 2021a is used to establish the FIS system, as illustrated in Fig. 3.14.

According to the monitoring results, the machine temperature is 93 °C and the noise level is 67 dB. After applying the FIS, the estimated time to the next failure is 220 h, as illustrated in Fig. 3.15.

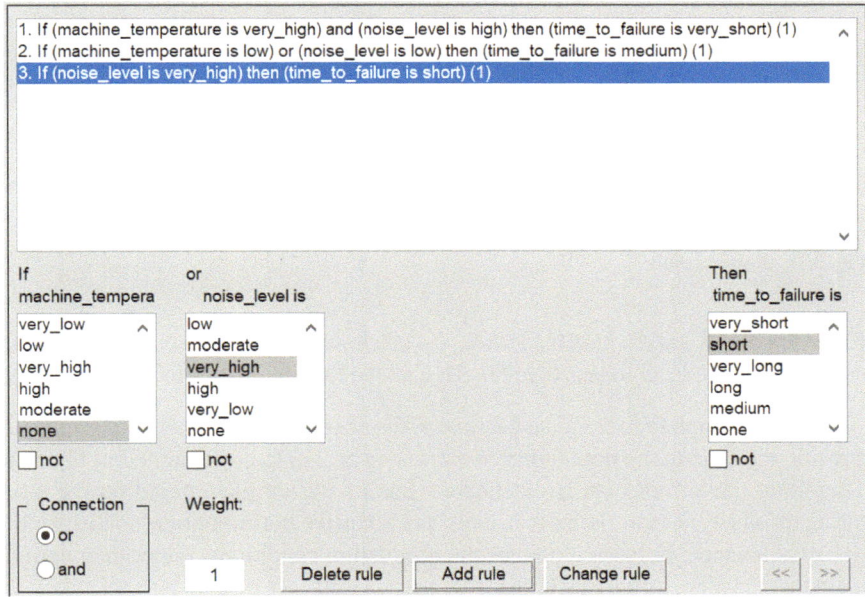

Fig. 3.14 FIS established using MATLAB 2021a

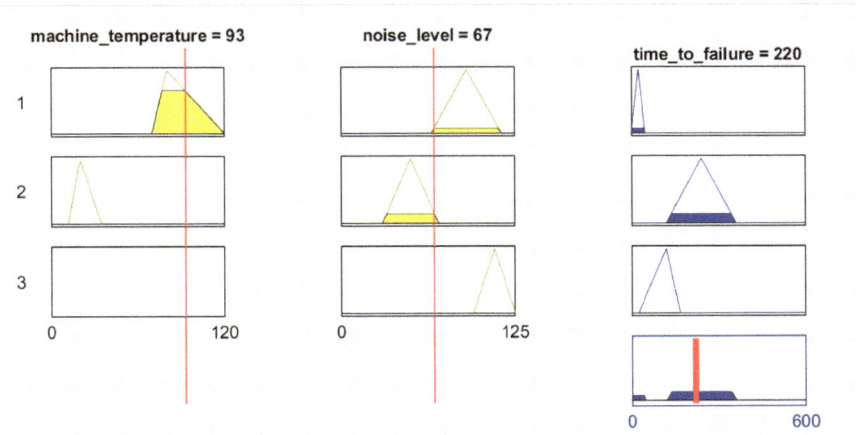

Fig. 3.15 Estimation result using the FIS

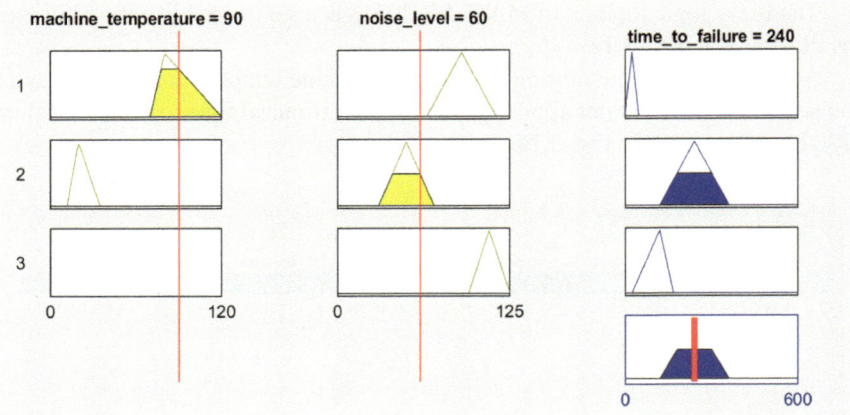

Fig. 3.16 Kaizen result

Example 3.4 A kaizen plan is to lower the machine temperature to below 90 °C and reduce the noise level to less than 60 dB. Can the time to next failure be postponed?

The FIS is applied to estimate the time to the next failure given the kaizen plan. As a result, the time to the next failure is estimated as 240 h, as illustrated in Fig. 3.16. Therefore, the kaizen plan is practicable. Such a kaizen plan, based on the monitoring results by sensors, helps to achieve **prescriptive maintenance**, not just predictive maintenance, because it adjusts the production conditions, rather than periodic maintenance plan, of the machine.

A response surface plot is drawn for the previous example in Fig. 3.17, showing that

- The relationship between the machine temperature and noise level and the time to the next failure is complex.
- It is possible to postpone the time to the next failure to 300 h.
- Conversely, the time to the next failure may be shorter than 50 h.

3.5   Cycle Time Reduction

Reducing the cycle time to manufacture a product has the following benefits:

- A shorter cycle time mean faster deliveries to customers. In other words, customers can be promised closer due dates [18], which helps win their orders.
- If the cycle times of all products are shorter, the level of work-in-process (WIP) in the factory will be lower [19], which is conducive to the management of the working environment.
- Shorter cycle times allow the factory to be better able to respond to emergency orders.

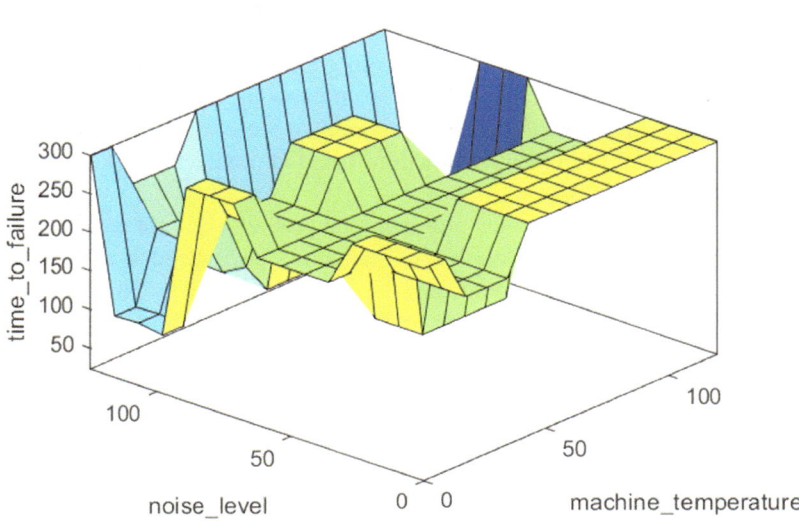

Fig. 3.17 Response surface plot

- Shorter cycle times reduce the variability of processing times and product quality encountered in the production process.

Therefore, shortening the cycle time of a product is the goal of every manufacturing system, especially a lean manufacturing system. There are multiple ways to achieve this.

Job scheduling optimization is the most frequently discussed method. For small manufacturing systems, some simple scheduling rules such as shortest processing time first (SPT), shortest remaining processing time first (SRPT), and weighted SPT (WSPT) can effectively reduce the average cycle time and the weighted average cycle time, respectively [20].

For large manufacturing systems, more complex scheduling rules must be applied. These scheduling rules may have parameters that must be estimated in advance. For example, Lu et al. [21] proposed the fluctuation smoothing policy for mean cycle time (FSMCT) for job scheduling in a wafer fabrication factory (wafer fab), in which the remaining cycle time of a job (i.e., the time from an intermediate step to the end of production) needs to be estimated in advance. To this end, AI technologies are helpful. For example, Chen [22] proposed a post-classifying fuzzy-neural approach to estimate the remaining cycle time of a job in a wafer fab, in which a fuzzy back propagation network (FBPN) was first constructed to estimate the remaining cycle time of a job based on its attributes. Subsequently, jobs were divided into clusters based on their estimation errors. A FBPN was then constructed for each cluster to estimate the remaining cycle times of jobs in the cluster.

Another way to schedule jobs in a complex manufacturing system is to formulate the scheduling problem as a mathematical programming model. For example, Grigoriev and Uetz [23] proposed a nonlinear programming (NLP) model to schedule jobs in a manufacturing system, in which the processing time of a job depended on the number of workers participating in the job: the more workers used, the shorter the processing time. The objective function was to minimize the makespan, i.e., the longest cycle time.

Sometimes such mathematical programming problems are not easy to solve. To overcome this difficulty, an evolutionary computing method, such as genetic algorithm (GA) [24], ant colony optimization [25], particle swarm optimization (PSO) [26], etc., can be applied to search for the global optimal solution.

Pull production is a popular technique in lean manufacturing for scheduling job operations, in which a job is pulled to the next workstation only when the workstation is available, eliminating the waiting time and thus reducing the cycle time [27].

It is not uncommon for the processing time of a step to be uncertain, especially for steps that require a human operator to operate the machine. In this case, the processing time can be modeled as a fuzzy number, and **fuzzy arithmetic** [5] is applied to perform the calculations required for pull production.

Implementing pull production becomes very challenging when there are many products with different processing times at each step. This challenge can be addressed by formulating the scheduling problem as a mathematical programming model [28].

Another method is through **lot sizing**, i.e., reducing the size of a job (or batch), which is also one of the methods often mentioned in lean manufacturing [29]. In theory, smaller jobs can flow faster, resulting in shorter cycle times. However, smaller jobs represent more jobs, which may require more frequent transportation and machine setups. These may waste a lot of capacity. In the literature, various AI techniques have been applied to determine the optimal size of a job.

References

1. S. Al Smadi, Kaizen strategy and the drive for competitiveness: challenges and opportunities. Compet. Rev. Int. Bus. J. **19**(3), 203–211 (2009)
2. A. Susilawati, J. Tan, D. Bell, M. Sarwar, Fuzzy logic based method to measure degree of lean activity in manufacturing industry. J. Manuf. Syst. **34**, 1–11 (2015)
3. K.E.K. Vimal, S. Vinodh, Application of artificial neural network for fuzzy logic based leanness assessment. J. Manuf. Technol. Manag. **24**(2), 274–292 (2013)
4. E. Akyar, H. Akyar, S.A. Düzce, A new method for ranking triangular fuzzy numbers. Int. J. Uncertain. Fuzziness Knowl-Based Syst. **20**(05), 729–740 (2012)
5. M. Hanss, *Applied Fuzzy Arithmetic* (Springer-Verlag, 2005)
6. E. Van Broekhoven, B. De Baets, Fast and accurate center of gravity defuzzification of fuzzy system outputs defined on trapezoidal fuzzy partitions. Fuzzy Sets Syst. **157**(7), 904–918 (2006)
7. J. Michalska, D. Szewieczek, The 5S methodology as a tool for improving the organization. J. Achiev. Mater. Manuf. Eng. **24**(2), 211–214 (2007)
8. J.S. Randhawa, I.S. Ahuja, An approach for justification of success 5S program in manufacturing organisations using fuzzy-based simulation model. Int. J. Prod. Qual. Manag. **25**(3), 331–348 (2018)

9. S. Abbasbandy, T. Hajjari, A new approach for ranking of trapezoidal fuzzy numbers. Comput. Math. Appl. **57**(3), 413–419 (2009)
10. E. Pourjavad, R.V. Mayorga, A comparative study and measuring performance of manufacturing systems with Mamdani fuzzy inference system. J. Intell. Manuf. **30**(3), 1085–1097 (2019)
11. T. Allahviranloo, R. Saneifard, Defuzzification method for ranking fuzzy numbers based on center of gravity. Iran. J. Fuzzy Syst. **9**(6), 57–67 (2012)
12. H. Shakouri, R. Nadimi, S.F. Ghaderi, Investigation on objective function and assessment rule in fuzzy regressions based on equality possibility, fuzzy union and intersection concepts. Comput. Ind. Eng. **110**, 207–215 (2017)
13. T. Chen, Y.C. Lin, A fuzzy-neural system incorporating unequally important expert opinions for semiconductor yield forecasting. Int. J. Uncertain. Fuzziness Knowl-Based Syst. **16**(01), 35–58 (2008)
14. H.C. Wu, T. Chen, C.H. Huang, A piecewise linear FGM approach for efficient and accurate FAHP analysis: smart backpack design as an example. Mathematics **8**(8), 1319 (2020)
15. T. Hafeez, L. Xu, G. Mcardle, Edge intelligence for data handling and predictive maintenance in IIOT. IEEE Access **9**, 49355–49371 (2021)
16. X. Chen, S. Jia, Y. Xiang, A review: knowledge reasoning over knowledge graph. Expert Syst. Appl. **141**, 112948 (2020)
17. C. Kahraman, D. Ruan, I. Doğan, Fuzzy group decision-making for facility location selection. Inf. Sci. **157**, 135–153 (2003)
18. P.C. Chang, J.C. Hsieh, T.W. Liao, A case-based reasoning approach for due-date assignment in a wafer fabrication factory, in *International Conference on Case-Based Reasoning* (2001), pp. 648–659.
19. J.D. Little, OR FORUM—Little's Law as viewed on its 50th anniversary. Oper. Res. **59**(3), 536–549 (2011)
20. M.L. Pinedo, *Scheduling: Theory, Algorithms, and Systems* (Springer, 2012)
21. S.C. Lu, D. Ramaswamy, P.R. Kumar, Efficient scheduling policies to reduce mean and variance of cycle-time in semiconductor manufacturing plants. IEEE Trans. Semicond. Manuf. **1**(3), 374–385 (1998)
22. T. Chen, Job remaining cycle time estimation with a post-classifying fuzzy-neural approach in a wafer fabrication plant: A simulation study. Proc. Inst. Mech. Eng. Part B J. Eng. Manuf. **223**(8), 1021–1031 (2009)
23. A. Grigoriev, M. Uetz, Scheduling jobs with time-resource tradeoff via nonlinear programming. Discret. Optim. **6**(4), 414–419 (2009)
24. F. Pezzella, G. Morganti, G. Ciaschetti, A genetic algorithm for the flexible job-shop scheduling problem. Comput. Oper. Res. **35**(10), 3202–3212 (2008)
25. C. Blum, M. Sampels, An ant colony optimization algorithm for shop scheduling problems. J. Math. Model. Algorithms **3**(3), 285–308 (2004)
26. C.J. Liao, C.T. Tseng, P. Luarn, A discrete version of particle swarm optimization for flowshop scheduling problems. Comput. Oper. Res. **34**(10), 3099–3111 (2007)
27. X.A. Koufteros, Testing a model of pull production: a paradigm for manufacturing research using structural equation modeling. J. Oper. Manag. **17**(4), 467–488 (1999)
28. N. Watanabe, S. Hiraki, A mathematical programming model for a pull type ordering system including lot production processes. Int. J. Oper. Prod. Manag. **15**(9), 44–58 (1995)
29. F. Zhou, P. Ma, Y. He, S. Pratap, P. Yu, B. Yang, Lean production of ship-pipe parts based on lot-sizing optimization and PFB control strategy. Kybernetes **50**(5), 1483–1505 (2020)

Chapter 4
AI Applications to Pull Production, JIT, and Production Leveling

4.1 Introduction

In a lean manufacturing system, in order to reduce the inventory of finished products (or work-in-process, WIP), **pull production** and **just-in-time (JIT)** are two important techniques. Pull production is to pull a job from an upstream workstation for processing only when a downstream workstation is about to become idle [1]. JIT is sending products (or WIP) to customers (or downstream workstations) just when they need it [2]. Pull production is undoubtedly a key way to achieve JIT. However, the following issues may hinder the implementation of JIT:

- **Unbalanced workstation capacity**: If some workstations are unable to meet the takt time, it will be difficult to deliver the order just in time.
- **Product quality problems**: If products are likely to be of poor quality, they have to be reworked, or more products need to be manufactured to make up for the shortage, delaying the delivery of the order.
- **Unstable processing**: If the processing time of each workstation is unstable, it is difficult to control the final completion time to reach JIT.

as illustrated in Fig. 4.1.

4.2 AI Applications to Pull Production

4.2.1 Fuzzy Logic for Pull Production Under Uncertainty

The operation at a workstation is usually completed by the operator and machine jointly, so the processing time is uncertain. To model such uncertainty, traditional probability statistics may not be easy to use because many assumptions must be met.

T.-C. T. Chen and Y.-C. Wang, *Artificial Intelligence and Lean Manufacturing*, SpringerBriefs in Applied Sciences and Technology, https://doi.org/10.1007/978-3-031-04583-7_4

Fig. 4.1 Factors contributing to JIT

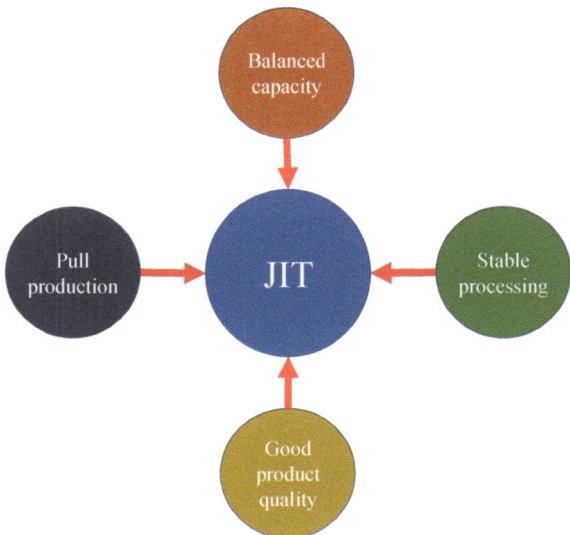

In contrast, **fuzzy logic** [3] is more convenient to apply. For example, the processing time at a workstation can be approximated by a triangular fuzzy number (TFN) [4] that has been introduced in Chap. 3.

Example 4.1 Pull production [1, 5] is a typical production control mode of lean manufacturing. Figure 4.2 is a simple pull production system of two workstations. The processing time at each workstation is approximated by a TFN. If 10 units of products are to be delivered at 18:00, the start and completion times of the operation at each workstation are calculated using fuzzy arithmetic as follows:

(Workstation #2)

Completion time = 18:00

Start time = 18:00 – 10 · (15, 22, 30) = 18:00 – (150, 220, 300) = (13:00, 14:20, 15:30)

The start time at Workstation #2 can be considered as the due date at Workstation #1.

(Workstation #1)

Completion time = (13:00, 14:20, 15:30) (i.e., the start time at Workstation #2)

Start time = (13:00, 14:20, 15:30) (–) 10 · (5, 7, 10) = (13:00, 14:20, 15:30) (–) (50, 70, 100) = (11:20, 13:10, 14:40)

Fig. 4.2 A simple pull production system of two workstations

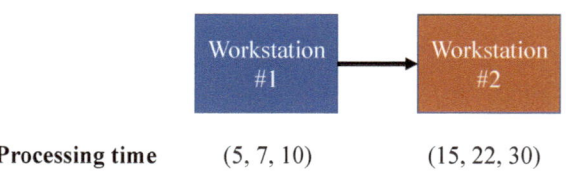

Table 4.1 Fuzzy production plan for Example 4.1

Start time	Completion time	Workstation
(11:20, 13:10, 14:40)	(13:00, 14:20, 15:30)	#1
(13:00, 14:20, 15:30)	18:00	#2

Table 4.2 Data of three orders

Order #	Quantity	Due date
1	2	18:00
2	1	19:30
3	5	17:50

If Workstation #1 starts production before 11:20, products will be made too early, resulting in unnecessary inventory; on the other hand, if the workstation starts production after 14:40, it will be too late for in-time delivery. The fuzzy production plan is shown in Table 4.1.

In the case of multiple orders, these orders can be arranged according to their due dates. That is, the latest order is arranged first. Then, the second latest order is scheduled before its due date and the start time of the latest order, similar to the calculation in project evaluation and review technique (PERT) [6], and so on.

Example 4.2 The data of three orders are summarized in Table 4.2. The scheduling task starts from Workstation #2. Order #2 has the latest due date and is arranged first:

(Order #2, Workstation #2)

Completion time = 19:30

Start time = $19:30 - 1 \cdot (15, 22, 30) = (19:00, 19:08, 19:15)$

Subsequently, Order #1 is scheduled before its due date and the start time of Order #2 as.

(Order #1, Workstation #2)

Completion time = Min(18:00, (19:00, 19:08, 19:15)) = 18:00.

Start time = $18:00 - 2 \cdot (15, 22, 30) = (17:00, 17:16, 17:30)$

At last, Order #3 is scheduled on this workstation as.

(Order #3, Workstation #2)

Completion time = Min(17:50, (17:00, 17:16, 17:30)) = (17:00, 17:16, 17:30)

Start time = $(17:00, 17:16, 17:30) (-) 5 \cdot (15, 22, 30) = (14:30, 15:26, 16:15)$

Subsequently, the operations at Workstation #1 are scheduled in the same way and in the same sequence:

(Order #2, Workstation #1)

Completion time = (19:00, 19:08, 19:15)

Start time = $(19:00, 19:08, 19:15) - 1 \cdot (5, 7, 10) = (18:50, 19:01, 19:10)$

(Order #1, Workstation #1)

Completion time = Min((17:00, 17:16, 17:30), (18:50, 19:01, 19:10)) = (17:00, 17:16, 17:30)

Start time = $(17:00, 17:16, 17:30) - 2 \cdot (5, 7, 10) = (16:40, 17:02, 17:20)$

(Order #3, Workstation #1)

Table 4.3 Fuzzy production plan for Example 4.2

Workstation #	Order #	Start time	Completion time
1	3	(13:40, 14:51, 15:50)	(14:30, 15:26, 16:15)
1	1	(16:40, 17:02, 17:20)	(17:00, 17:16, 17:30)
1	2	(18:50, 19:01, 19:10)	(19:00, 19:08, 19:15)
2	3	(14:30, 15:26, 16:15)	17:50
2	1	(17:00, 17:16, 17:30)	18:00
2	2	(19:00, 19:08, 19:15)	19:30

Completion time = Min((14:30, 15:26, 16:15), (16:40, 17:02, 17:20)) = (14:30, 15:26, 16:15)

Start time = (14:30, 15:26, 16:15) (−) 5 · (5, 7, 10) = (13:40, 14:51, 15:50)

Finally, the fuzzy pull production plan is presented in Table 4.3.

The start and completion times of an operation are expressed as ranges, giving the operator a considerable degree of flexibility. However, although JIT can achieve the effect of zero inventory, it may lead to idle (or wasted) production capacity, as shown in Fig. 4.3. This problem can be solved if some inventory is allowed and orders are fulfilled in a timely manner rather than JIT [7].

It is common that not all products are of good quality. The yield of a product is the percentage of workpieces that are of acceptable quality after a step (or all steps), which needs to be considered in planning operations.

Example 4.3 In Example 4.1, the yield of the product is about 90%, which is modeled with a TFN as (82%, 90%, 95%). To generate 10 units of acceptable products, the

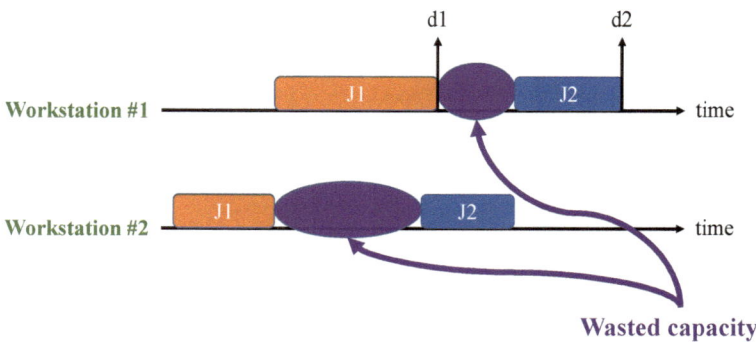

Fig. 4.3 Capacity waste caused by pull production

number of workpieces that need to be manufactured is 10/(82%, 90%, 95%) = (10.5, 11.1, 12.2) → (11, 11, 12) after rounding to integers. Therefore, the start and completion times for the operation at each workstation are calculated as.

(Workstation #2)

Completion time = 18:00

Start time = 18:00 − (11, 11, 12) (×) (15, 22, 30) = 18:00 − (165, 242, 360) = (12:00, 13:58, 15:15)

(Workstation #1)

Completion time = (12:00, 13:58, 15:15)

Start time = (12:00, 13:58, 15:15) (−) (11, 11, 12) (×) (5, 7, 10) = (12:00, 13:58, 15:15) (−) (55, 77, 120) = (10:00, 12:41, 14:20)

Obviously, the operation at each workstation must be brought forward after taking into account quality loss, as illustrated in Fig. 4.4.

Example 4.4 The problem of quality loss is considered for Example 4.2. The quantities of workpieces that need to be manufactured for the three orders are

Order #1: 2 / (82%, 90%, 95%) = (2.11, 2.22, 2.44) → (2, 2, 2)

Order #2: 1 / (82%, 90%, 95%) = (1.05, 1.11, 1.22) → (1, 1, 1)

Order #3: 5 / (82%, 90%, 95%) = (5.26, 5.56, 6.10) → (5, 6, 6)

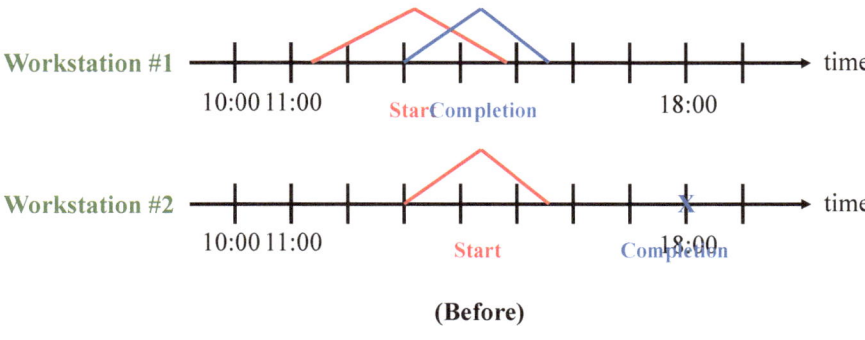

(Before)

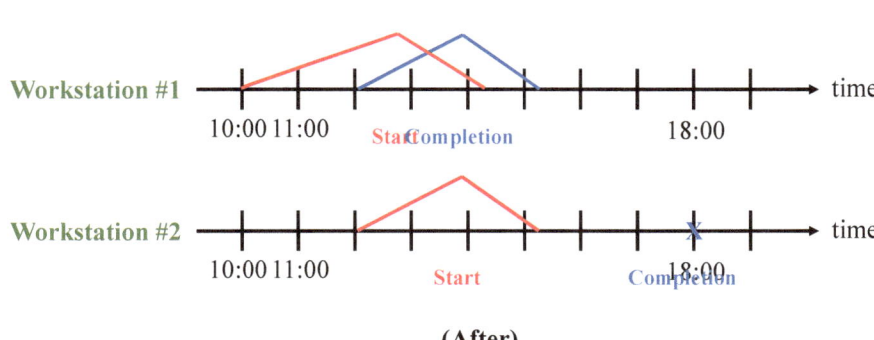

(After)

Fig. 4.4 Comparison of JIT production plans before and after considered quality loss

Table 4.4 Fuzzy production plan for Example 4.4

Workstation #	Order #	Start time	Completion time
1	3	(13:00, 14:22, 15:50)	(14:00, 15:04, 16:15)
1	1	(16:40, 17:02, 17:20)	(17:00, 17:16, 17:30)
1	2	(18:50, 19:01, 19:10)	(19:00, 19:08, 19:15)
2	3	(14:00, 15:04, 16:15)	17:50
2	1	(17:00, 17:16, 17:30)	18:00
2	2	(19:00, 19:08, 19:15)	19:30

The new fuzzy production plan is shown in Table 4.4.

If there are both pull production (orders with due dates) and push production (orders without due dates, but should be completed the sooner the better) in a manufacturing system, orders for pull production should be scheduled first. Then, when the production capacity is idle, orders for push production are placed.

4.2.2   Artificial Neural Networks for Cycle Time Estimation and Job Scheduling in Pull Production

The pull production of a large manufacturing system is even more complicated. A trend in practice is to adopt more real-time manufacturing execution system (MES) and production management information system (PROMIS) [8]. These systems rely on the applications of AI. For example, a large manufacturing system is illustrated in Fig. 4.5. It is difficult to perform the above-mentioned calculations. Instead, the cycle time (i.e., the output time subtracting the input time) of a workpiece can be estimated.

The problem is how to estimate the cycle time of a workpiece. If the manufacturing system is in a stable state for a long time, the average cycle time of old workpieces can be used as a reference value.

Example 4.5 The processing time at workstation M1 is 13 min, while the cycle time of a workpiece from this workstation to completion is estimated as 167 min. If 25 units of products are to be delivered at 18:00, then the first workpiece should be input to workstation M1 no later than $18:00 - 167 - (25 - 1) \cdot 13 = 10:01$, as illustrated in Fig. 4.6.

Otherwise, an AI method can be applied to estimate the cycle time of a workpiece. For example, Chen [9] estimated the cycle time of a job (including multiple

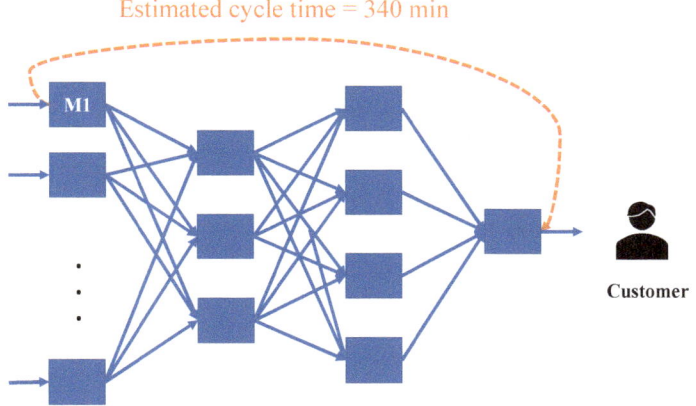

Fig. 4.5   A large manufacturing system

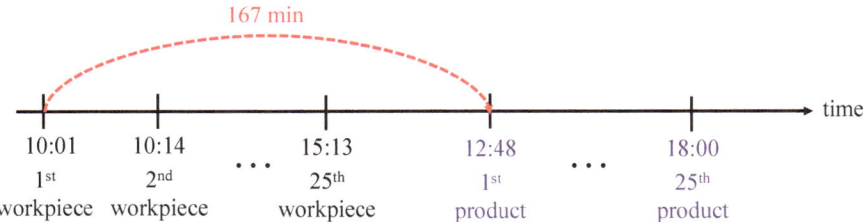

Fig. 4.6   Pull production plan for Example 4.5

workpieces) by considering the values of the following production conditions when the job was released into the factory:

- Factory utilization: the highest utilization of all machines in the factory on the previous day.
- Job size: the number of workpieces that are processed together.
- Queue length before the bottleneck;
- Factory WIP;
- Average cycle time of recently completed jobs;
- Factory queue.

Then, a **backpropagation neural network (BPN)** (or **feedforward neural network, FNN**) was constructed to estimate the cycle time from the values of these production conditions, as illustrated in Fig. 4.7.

In Fig. 4.7, the BPN has three layers: the input layer, a hidden layer with 12 nodes, and the output layer. It is a shallow artificial neural network (ANN) since there is only a single hidden layer. The values of the six production conditions, indicated by $x_{j1} \sim x_{j6}$, are inputted into the BPN, which are transmitted to the hidden layer as follows:

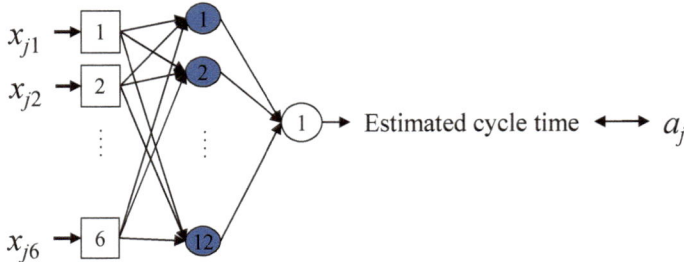

Fig. 4.7 BPN for estimating the cycle time of a job

$$h_{jl} = \frac{1}{1 + e^{-n_{jl}^h}} \tag{4.1}$$

where

$$n_{jl}^h = I_{jl}^h - \theta_l^h \tag{4.2}$$

$$I_{jl}^h = \sum_{p=1}^{6} w_{pl}^h x_{jp} \tag{4.3}$$

where h_{jl} is the output from node l of the first hidden layer; $l = 1 \sim 12$. θ_l^h is the threshold on this node; w_{pl}^h is the connection weight between input node p and this node. h_{jl} is passed to the output layer. Then, the network output o_j is generated as

$$o_j = n_j^o \tag{4.4}$$

where

$$n_j^o = I_j^o - \theta^o \tag{4.5}$$

$$I_j^o = \sum_{l=1}^{12} w_l^o h_{jl} \tag{4.6}$$

The network output is the estimated cycle time that is compared with the cycle time (i.e., actual value) of the job a_j.

In theory, many algorithms can be applied to train the BPN to derive the optimal values of network parameters, e.g., the gradient descent (GD) algorithm, the conjugate gradient algorithm, the scaled conjugate gradient algorithm, Levenberg–Marquardt (LM) algorithm, etc. In practice, software or programming languages that support ANN applications are not only common but also mature.

An illustrative example is given below.

Example 4.6 Cycle time related data have been collected for 120 jobs and are presented in Table 4.5. The data of the first 80 jobs are used to train the BPN, while the remaining is reserved for evaluating the estimation performance in terms of root mean squared error (RMSE):

$$RMSE = \sqrt{\frac{\sum_{j=1}^{n} (o_j - a_j)^2}{n}} \tag{4.7}$$

The MATLAB program for implementing the BPN is shown in Fig. 4.8. The BPN is trained using the Levenberg–Marquardt (LM) algorithm [10]. The number of epochs is 50000.

The training process is illustrated in Fig. 4.9. The forecasting results are summarized in Fig. 4.10. The forecasting accuracy, measured in terms of RMSE for test data, is 198 h.

Table 4.5 Cycle time-related data of 120 jobs

j	x_{j1} (%)	x_{j2}	x_{j3}	x_{j4}	x_{j5}	x_{j6}	a_j
1	84.20	24	99	807	1223	158	953
2	94.80	23	142	665	1225	164	1248
...							
80	91.90	24	232	807	1333	166	1272
81	85.20	24	127	785	1251	182	1173
...							
120	88.80	22	326	777	1319	159	1285

```
training_x=[84.20% 94.80% ... 91.90%;24 23 ... 24;99 142 ... 232;807 665 ... 807;1223 1225 ...
1333;158 164 ... 166];
test_x=[85.20% ... 88.80%;24 ... 22;127 ... 326;785 ... 777;1251 ... 1319;182 ... 159];
training_y=[953 1248 ... 1272];
test_y=[1173 ... 1285];
net=feedforwardnet(12);
net.dividefcn='dividetrain';
net.trainParam.lr=0.1;
net.trainParam.epochs=50000;
net.trainParam.goal=100;
net=train(net,training_x,training_y);
training_est_ct=net(training_x);
test_est_ct=net(test_x);
test_rmse=mean((test_y-test_est_ct).^2)^0.5;
```

Fig. 4.8 MATLAB code for implementing the BPN

Fig. 4.9 Training process

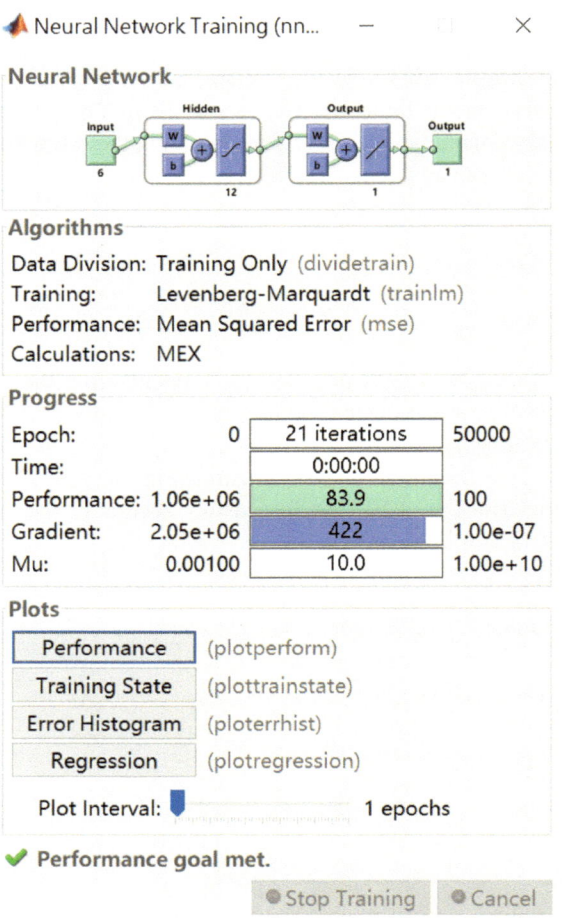

As shown by this example, the application of ANNs is quite simple. For lean manufacturing practitioners, ANNs can be considered as black boxes that model nonlinear cause-and-effect relationships effectively.

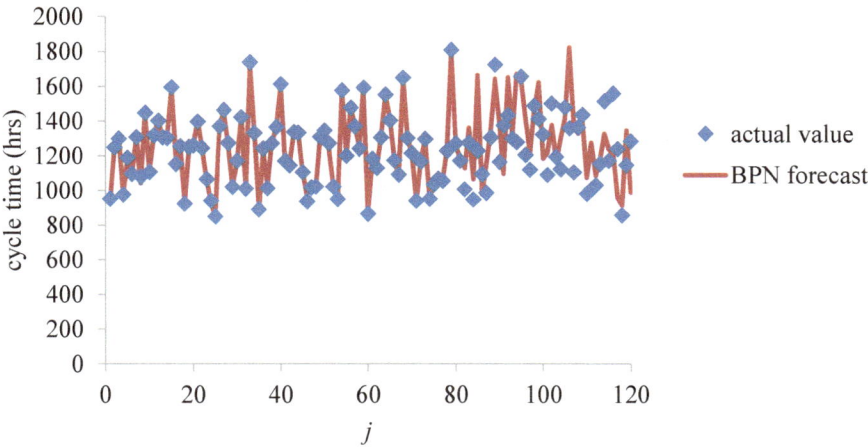

Fig. 4.10 Forecasting results

4.3 AI Applications to JIT

4.3.1 3D Printing Applications to Lean Manufacturing

The goal of lean manufacturing is waste reduction. 3D printing can make products in the most accurate way, so it can save the consumption of materials and avoid unnecessary wastes. From this point of view, 3D printing can indeed lead to lean manufacturing. However, the relationship between 3D printing and lean manufacturing has rarely been investigated.

In the view of Campbell et al. [11], 3D printing reduces or eliminates the need for multiple raw materials, leading to the disappearance of associated assembly lines and supply chains. As a result, the entire industry has become leaner.

3D printing can also be applied to make the parts of a product. Calì et al. [12] used 3D printing to make the posable joints of an animal model, and then established an efficient procedure to fit these posable joints by considering rotational constraints, referring to several concepts of lean manufacturing.

Chen and Lin [13] mentioned that 3D printing can contribute to lean manufacturing in many ways, as shown in Fig. 4.11.

The most direct benefit of 3D printing is small batches. All 3D printed objects may be different, achieving mass customization. With 3D printing, products can be manufactured in a print-on-demand manner, eliminating the necessity of building an inventory for the product, in line with the concepts of "pull system" and "no inventory" in lean manufacturing.

The separation of man and machine is also an important principle of lean manufacturing. It shifts the tasks of workers from simple machine operations to the participation in a continuous improvement process [14]. The operations of a 3D printer

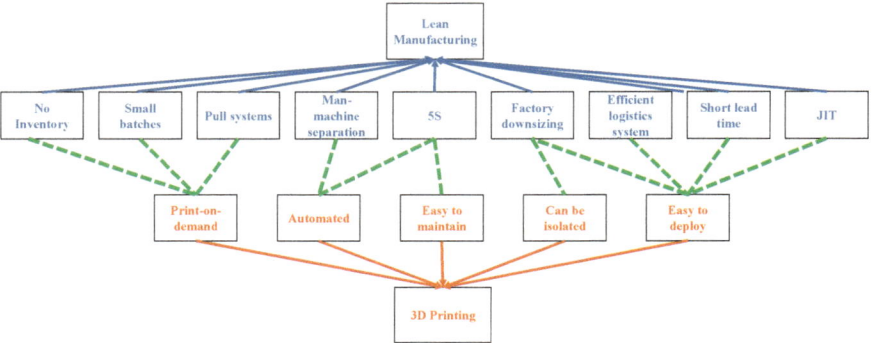

Fig. 4.11 Contributions of 3D printing to lean manufacturing

are largely automated. This property facilitates the separation of man and machine. In addition, 3D printers are easy to maintain, which is good for 5S implementation.

Manufacturing facilities equipped with 3D printers can be isolated from other facilities, allowing factories to be downsized and made leaner. According to the philosophy of lean manufacturing, wastes are usually caused by excessive or uneven workloads; therefore, a manufacturing facility composed of 3D printers can distribute the workload so that the burden is evenly shared by the 3D printers, which is also in line with this philosophy [15].

4.3.2 3D Printing and JIT

In a ubiquitous manufacturing (UM) network of 3D printers, customers can choose the nearest 3D printer to print products, thereby shortening the lead time, distance, and cost of delivering products. In sum, the efficiency of the logistics system is improved. Furthermore, JIT manufacturing is a key objective of lean manufacturing by considering the location of the customer, the due date of the order, and the required printing and shipping time (depending on the location of the customer).

In the literature, there have been some studies that constructed UM systems consisting of 3D printing facilities to achieve JIT fulfillment of customer orders. For example, Chen and Lin [13] conducted a literature review to examine the feasibility of collaborating multiple 3D printing facilities on order fulfillment. They also discussed how to optimize the overall performance of these 3D printing facilities.

In Chen and Wang [16], customers on the move placed their orders for 3D objects through smart phones. These orders were distributed among multiple 3D printing facilities. A mixed integer-quadratic programming (MIQP) problem was solved to balance the loads on 3D printing facilities and plan the shortest delivery path through these 3D printing facilities. A branch-and-bound algorithm was also proposed to help solve the MIQP problem.

Following their study, Wang et al. [17] derived the slack for each 3D printing facility, so that a 3D object could be reprinted if the printing process was early terminated before the slack.

In existing models, the transportation vehicle departs for a 3D printing facility while the printing process is still in progress to avoid waiting. As a result, orders are usually distributed to distant 3D printing facilities, resulting in unbalanced loads on 3D printing facilities. To address this issue, Chen and Lin [18] evaluated the suitability of a 3D printing facility in terms of the closeness to JIT.

The following optimization problem can be solved to choose the suitable facility for printing a 3D object in a JIT manner:

$$\text{Min } Z_1 = w_k \tag{4.8}$$

subject to

$$w_k = t_c + p - t_s - d_k \tag{4.9}$$

where w_k is the time that the customer needs to wait after arriving at the k-th 3D printing facility. Only 3D printing facilities near the customer are considered. t_c is the current time, p is the time required for printing the 3D object, t_s is the time the customer leaves for the 3D printing facility, and d_k is the minimum time required to reach the k-th 3D-printing facility. The objective function is to minimize the waiting time, meaning JIT. This model is a linear programming model that can be solved using an existing optimization software.

The value of d_k is derived in either of the following two ways:

- Solve the following mixed integer-nonlinear programming (MINLP) problem:

$$\text{Min } Z_2 = d_k \tag{4.10}$$

subject to

$$d_i \leq d_j + l_{ji}, \ i = 0 \sim k; \ l_{ij} \neq \infty \tag{4.11}$$

$$d_i = \sum_{j < i, l_{ji} \neq \infty} x_{ji}(d_j + l_{ji}), \ i = 0 \sim k \tag{4.12}$$

$$\sum_{j < i, l_{ji} \neq \infty} x_{ji} = 1, \ i = 0 \sim k \tag{4.13}$$

$$x_{ji} \in \{0, 1\}, i = 0 \sim k; j < i; \ l_{ij} \neq \infty \tag{4.14}$$

where l_{ij} is the length of the shortest path between locations i and j.

Fig. 4.12 Closeness to JIT

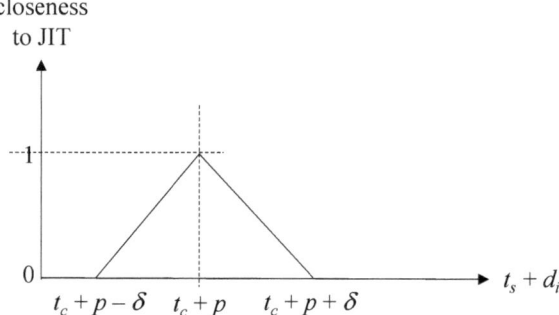

- Apply a mobile navigation service such as Google Maps

Lin and Chen [19] proposed a modified Dijkstra algorithm to solve the 3D printing facility selection problem, in which the suitability of a 3D printing facility was evaluated as

$$s_i = 1 - \frac{t_c + p}{t_s + d_i} \qquad (4.15)$$

The most suitable 3D printing facility can thus be determined. However, (4.12) tends to direct a customer to a more distant facility to minimize the waiting time [20]. For this reason, Chen and Lin defined the closeness to JIT (see Fig. 4.12) to evaluate the suitability of a 3D printing facility:

$$s_i = \min\left(\max\left(\frac{t_s + d_i - t_c - p + \delta}{\delta},\ 0\right),\ \max\left(\frac{-t_s - d_i + t_c + p + \delta}{\delta},\ 0\right)\right) \qquad (4.16)$$

where δ is the tolerance. According to this figure, if $t_s + d_k = t_c + p$, $w_k = 0$. The 3D object can be made in a JIT manner, and the customer does not arrive too early or need to wait.

Chen [21] defined the printing time with a fuzzy number to account for its uncertainty. Then, a fuzzy mixed integer-linear programming (FMILP) model and a fuzzy MIQP (FMIQP) model were optimized, respectively, to balance the loads on 3D printing facilities and plan the shortest delivery path.

Subsequently, if N 3D objects are to be made by K 3D printing facilities collaboratively, another MILP model can be optimized to make the production plan. Assuming a_k is the available time of 3D printing facility k. n_k is the number of 3D objects to be printed by 3D printing facility k. Then, all 3D objects that require 3D printing facility k to print can be done at $a_k + n_k p_k$. Letting the transportation time between 3D printing facility k and the customer be denoted by $\tilde{d}_k$. Then, the printed 3D objects can be transported to the customer at $a_k + n_k p_k + d_k$. All 3D printing facilities must transport 3D objects to the customer to fulfill the order, so the order fulfillment time is $\max_{k}(a_k + n_k p_k + d_k)$, which is to be minimized:

$$Min\, Z_3 = \max_{k}(a_k + n_k p_k + d_k) \tag{4.17}$$

subject to

$$\sum_{k=1}^{K} n_k = N \tag{4.18}$$

$$n_k \in Z^+ \cup \{0\}; k = 1 \sim K \tag{4.19}$$

Example 4.7 A customer placed an order for six 3D objects that will be made by three 3D printing facilities jointly. Therefore, $N = 6$ and $K = 3$. The available time of each 3D printing facility, the printing time of a 3D object at the 3D printing facility, and the transportation time from the 3D printing facility to the customer have been estimated and are summarized in Table 4.6.

The MILP model for making the production plan is formulated using Lingo, as illustrated in Fig. 4.13. The optimal solution is $\{n_k\} = \{2,\ 2,\ 2\}$, giving $Z_3^* = 493$ (min).

In fact, both the printing time and the transportation time are subject to uncertainty. After considering this, the following FMILP model can be optimized to make the production plan:

$$Min\, \tilde{Z}_4 = \max_{k}(a_k + n_k \tilde{p}_k (+) \tilde{d}_k) \tag{4.20}$$

subject to

Table 4.6 Data of three 3D printing facilities	k	a_k(min)	p_k(min)	d_k(min)
	1	0	195	14
	2	36	174	10
	3	15	231	16

```
min=Z3;
Z3>=0+n1*195+14;
Z3>=36+n2*174+10;
Z3>=15+n3*231+16;
n1+n2+n3=6;
@gin(n1);@gin(n2);@gin(n3);
```

Fig. 4.13 Lingo code for the MINLP problem

$$\sum_{k=1}^{K} n_k = N \tag{4.21}$$

$$n_k \in Z^+ \cup \{0\}; k = 1 \sim K \tag{4.22}$$

which can be converted into the following MILP problem:

$$Min \ Z_5 = \frac{Z_{41} + Z_{42} + Z_{43}}{3} \tag{4.23}$$

subject to

$$Z_{41} = \max_{k}(a_k + n_k p_{k1} + d_{k1}) \tag{4.24}$$

$$Z_{42} = \max_{k}(a_k + n_k p_{k2} + d_{k2}) \tag{4.25}$$

$$Z_{43} = \max_{k}(a_k + n_k p_{k3} + d_{k3}) \tag{4.26}$$

$$\sum_{k=1}^{K} n_k = N \tag{4.27}$$

$$n_k \in Z^+ \cup \{0\}; k = 1 \sim K \tag{4.28}$$

Most existing systems of 3D printing facilities are managed centrally. Vatankhah Barenji et al. [22] applied the blockchain technology to transform such systems, allowing 3D printing facilities to communicate with each other and manage the system by themselves.

4.4   Production Leveling

4.4.1   Production Leveling Based on TAKT Time

Production leveling, also known as production smoothing or heijunka, is a technique to eliminate Mura (unevenness), thereby reducing Muda (wastes). The goal is to produce intermediate products at a constant rate so that downstream processing can also proceed at a constant and predictable rate [23].

An important concept for production leveling is takt time, which is the rate at which a manufacturing system needs to generate a product to meet customer demand [24]. The takt time of a manufacturing system can be calculated as

$$\text{Takt time} = \frac{\text{total working time}}{\text{demand}} \tag{4.29}$$

For example, if the demand is 800 pieces per week, and the production system runs 5 days a week and 8 h a day, then the takt time is $(5 \cdot 8) / 800 = 0.05$ h, which means generating a product every 3 min. The takt time, often confused with the cycle time, but is different. The takt time varies with demand. Takt comes from German and means baton. Planning production according to the takt time allows the entire production line to work at the same rhythm like a symphony orchestra. However, the time required for each production step varies. How to get all the steps to produce according to the takt time is a challenging task.

Adjusting the processing time of each workstation to be equal to or slightly less than the takt time is the goal of production leveling. There are various approaches for this purpose:

- Operators of previous or next workstations help perform part of the work, which is based on a detailed work study [25].
- Merging two (or more) consecutive steps and then break down into several new steps, each with a production time that meets the takt time requirement.
- Increasing the number of machines.
- Improving production efficiency and shortening the processing time.

The approaches are different when the processing time is too long and when it is too short, as shown in Fig. 4.14.

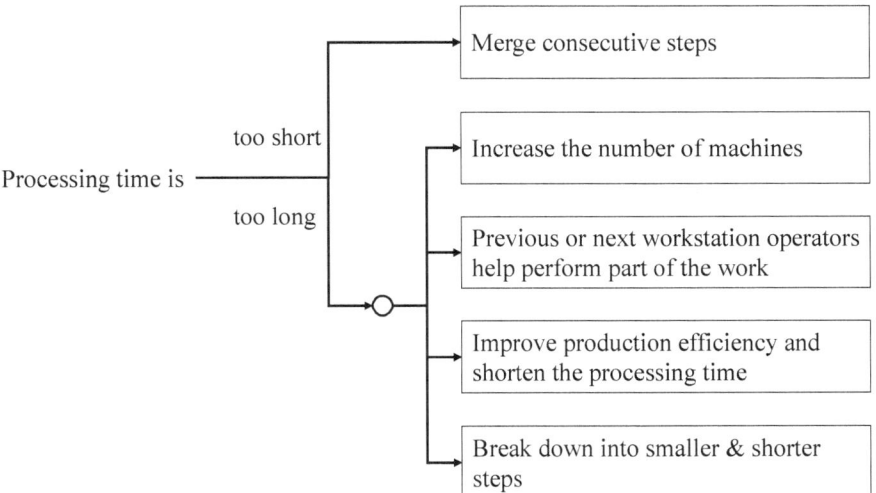

Fig. 4.14 Approaches for adjusting the processing time to match the takt time

4.4.2   Cloud Manufacturing Application to Production
Leveling

The emergence of cloud manufacturing (CMfg) technology provides a new opportunity for production leveling. CMfg is a manufacturing model that enables ubiquitous, convenient, and real-time access to a shared pool of configurable manufacturing resources over the Internet [19, 26]. CMfg is an application of cloud computing [27] in manufacturing to rapidly provision and release manufacturing resources with minimal administrative efforts or service provider interaction.

The capacity of a workstation is the number of workpieces that can be processed in a period of time, which can be calculated as follows:

$$\text{Capacity} = \frac{\text{total working time}}{\text{processing time}} \qquad (4.30)$$

The capacity of a step is inversely proportional to the processing time. In other words, the capacity of a step with a longer processing time is lower. In a lean manufacturing system, a processing time longer than the takt time means a lack of sufficient capacity. Conversely, a processing time shorter than the takt time means excess capacity.

CMfg can be applied to production leveling as follows. When the capacity of a step is insufficient, the factory can seek outside, cloud-based capacity through the intervention of a cloud service provider to make up. In contrast, when a step has excess capacity, it can also be provided to other factories in the same way as cloud-based capacity [28]. In other words, CMfg provides another way to match the takt time, as illustrated in Fig. 4.15.

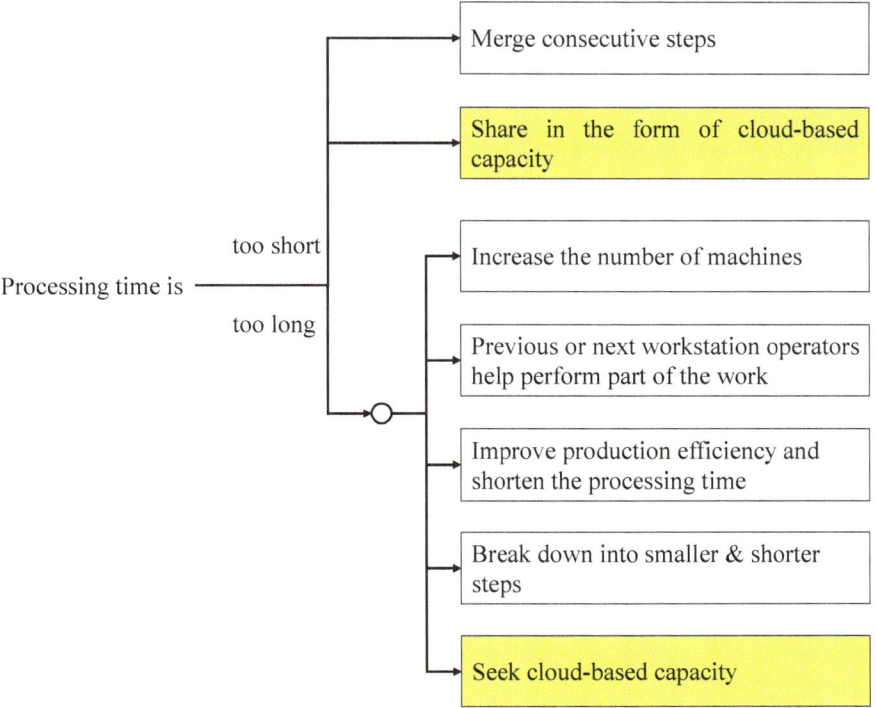

Fig. 4.15 Application of CMfg to production leveling

References

1. Lean Enterprise Institute, Pull production (2021). https://www.lean.org/lexicon-terms/pull-production/
2. G. Singh, I.S. Ahuja, Just-in-time manufacturing: literature review and directions. Int. J. Bus. Contin. Risk Manag. **3**(1), 57–98 (2012)
3. A. Susilawati, J. Tan, D. Bell, M. Sarwar, Fuzzy logic based method to measure degree of lean activity in manufacturing industry. J. Manuf. Syst. **34**, 1–11 (2015)
4. M. Hanss, *Applied Fuzzy Arithmetic* (Springer-Verlag, 2005)
5. X.A. Koufteros, Testing a model of pull production: a paradigm for manufacturing research using structural equation modeling. J. Oper. Manag. **17**(4), 467–488 (1999)
6. F. Habibi, O. Birgani, H. Koppelaar, S. Radenović, Using fuzzy logic to improve the project time and cost estimation based on Project Evaluation and Review Technique (PERT). J. Proj. Manag. **3**(4), 183–196 (2018)
7. M.L. Pinedo, *Scheduling: Theory, Algorithms, and Systems* (Prentice Hall, 2012)
8. P. Perico, J. Mattioli, Empowering process and control in lean 4.0 with artificial intelligence, in *Third International Conference on Artificial Intelligence for Industries* (2020), pp. 6–9
9. T. Chen, A hybrid SOM-BPN approach to lot output time prediction in a wafer fab. Neural Process. Lett. **24**(3), 271–288 (2006)
10. G. Lera, M. Pinzolas, Neighborhood based Levenberg–Marquardt algorithm for neural network training. IEEE Trans. Neural Netw. **13**(5), 1200–1203 (2002)
11. T. Campbell, C. Williams, O. Ivanova, B. Garrett, Could 3D printing change the world. Technol. Potential Implic. Addit. Manuf. **10**, 1–15 (2011)

12. J. Calì, D.A. Calian, C. Amati, R. Kleinberger, A. Steed, J. Kautz, T. Weyrich, 3D-printing of non-assembly, articulated models. ACM Trans. Graph. **31**(6), 130 (2012)
13. T. Chen, Y.C. Lin, Feasibility evaluation and optimization of a smart manufacturing system based on 3D printing: a review. Int. J. Intell. Syst. **32**(4), 394–413 (2017)
14. Velaction Continuous Improvement, Separate man from machine (2016). http://www.velaction.com/separate-man-from-machine/
15. A. Goel, M.K. Sundararajan, The flat supply chain (2007). http://www.sdcexec.com/article/10289663/the-flat-supply-chain
16. T. Chen, Y.C. Wang, An advanced fuzzy approach for modeling the yield improvement of making aircraft parts using 3D printing. Int. J. Adv. Manuf. Technol. **105**(10), 4085–4095 (2019)
17. Y.C. Wang, T. Chen, Y.L. Yeh, Advanced 3D printing technologies for the aircraft industry: a fuzzy systematic approach for assessing the critical factors. Int. J. Adv. Manuf. Technol. **105**(10), 4059–4069 (2019)
18. T.C.T. Chen, Y.C. Lin, A three-dimensional-printing-based agile and ubiquitous additive manufacturing system. Robot. Comput.-Integr. Manuf. **55**, 88–95 (2019)
19. Y.-C. Lin, T. Chen, A ubiquitous manufacturing network system. Robot. Comput.-Integr. Manuf. **45**, 157–167 (2017)
20. T. Chen, Creating a just-in-time location-aware service using fuzzy logic. Appl. Spat. Anal. Policy **26**(9), 287–307 (2016)
21. T.C.T. Chen, Fuzzy approach for production planning by using a three-dimensional printing-based ubiquitous manufacturing system. AI EDAM **33**(4), 458–468 (2019)
22. A. Vatankhah Barenji, Z. Li, W.M. Wang, G.Q. Huang, D.A. Guerra-Zubiaga, Blockchain-based ubiquitous manufacturing: a secure and reliable cyber-physical system. Int. J. Prod. Res. **58**(7), 2200–2221 (2020)
23. D.T. Jones, Heijunka: Leveling production (2006). https://www.sme.org/technologies/articles/2006/heijunka-leveling-production/
24. Kanbanize.com, What is Takt time and how to define it? (2022). https://kanbanize.com/continuous-flow/takt-time
25. H. Elghobary, A. Haridi, M. Naguib, Computerized work study approach to factory design. Comput. Ind. Eng. **13**(1–4), 327–331 (1987)
26. X. Xu, From cloud computing to cloud manufacturing. Robot. Comput.-Integr. Manuf. **28**(1), 75–86 (2012)
27. F. Alharbi, A. Atkins, C. Stanier, Understanding the determinants of cloud computing adoption in Saudi healthcare organisations. Complex Intell. Syst. **2**(3), 155–171 (2016)
28. T. Chen, Y.-C. Wang, A fuzzy mid-term capacity and production planning model for a manufacturer under a cloud manufacturing environment. Complex Intell. Syst. **7**, 71–85 (2021)

Chapter 5
AI Applications to Shop Floor Management in Lean Manufacturing

5.1 Shop Floor Management

5.1.1 Introduction

Shop floor management encompasses specific activities on the production floor and elsewhere in the factory, with the aim of creating a clear, safe, stable, and actionable workflow [1]. However, there is no consensus as to what activities are included in shop floor management. According to Oracle.com [2], shop floor management includes the following activities.

- Hours and quantities tracking.
- Reporting.
- Material tracking.
- Manufacturing accounting.
- Production scheduling and tracking.
- Work order scheduling.
- Process or routing instructions.
- Part list generation.

Periodic maintenance and repair and quality control also belong to shop floor management.

5.1.2 Shop Floor Management in Lean Manufacturing

In principle, shop floor management methods and tools used in general manufacturing systems can also be applied to a lean manufacturing system. However, some methods

© The Author(s), under exclusive license to Springer Nature Switzerland AG 2022
T.-C. T. Chen and Y.-C. Wang, *Artificial Intelligence and Lean Manufacturing*,
SpringerBriefs in Applied Sciences and Technology,
https://doi.org/10.1007/978-3-031-04583-7_5

and tools have been proposed or designed for the shop floor management of a lean manufacturing system, such as.

- **Continuous flow**: Continuous flow is a smooth production method used to manufacture, produce, or process materials with fewest or even no buffers between steps [3, 4].
- **Pull production**: Pull production is a method of production control in which downstream activities signal their needs to upstream activities [4, 5]. AI applications to pull production have been introduced in Chap. 4.
- **Single minute exchange of die (SMED)**: SMED is a process for re-setting up production equipment from one product type to another as soon as possible [6, 7].
- **Total productive maintenance (TPM)**: TPM is a management philosophy that engages all employees in the improvement activities of eliminating the six wastes during the life cycle of equipment [8].
- **Total quality management (TQM)**: TQM is a management philosophy in which all departments, employees, and managers are responsible for continuously improving quality so that products and services meet or exceed customer expectations [9, 10].

5.1.3   Necessity of Artificial Intelligence Applications in Shop Floor Management

Undoubtedly, shop floor management involves precise measurement, calculation, and analysis of time, quantity, quality, and other data. For this purpose, the application of artificial intelligence (AI) technologies is necessary:

- Many data related to time, quantity, and quality have become **big data**, which involves datasets of sizes that exceed the capabilities of typical database software tools to capture, store, manage, and analyze [11]. To analyze these big data, the application of AI technologies will be more effective and efficient.
- Some specific data analyses are difficult for humans to do manually, and AI must be applied to assist.
- Only by collecting relevant data in real time can the accuracy of analysis results be ensured, for which the application of smart technologies, such as sensors and wireless communication, is crucial.
- Software related to AI applications is already common and not necessarily expensive. For example, the free R language can also be used to build **artificial neural networks (ANNs)** to analyze nonlinear relationships [12]. To survive in an increasingly competitive industry, there is no reason for businesses not to use them.

5.2 AI Applications to Shop Floor Management in Lean Manufacturing

5.2.1 Lean Data

One of the main tasks of big data analytics is to collect huge amounts of data. However, big data are difficult to analyze. Sampling, principal component analysis (PCA), and other techniques have traditionally been applied to reduce the amount of data used in the analysis. In contrast, Küfner et al. [13] mentioned the concept of **lean data**, meaning to reduce the amount of data that are collected before being transmitted to the user. For this purpose, they proposed the "decentralized data reduction" approach, in which an ANN analysis module was embedded in each machine to analyze the data generated by the machine in real time, and finally presented analysis results, rather than the original data, to the user.

Edge computing, or **edge intelligence**, developed in recent years, is a similar concept that enables sensors to process or analyze the collected data [14]. These techniques are for data reduction that is a common practice in dealing with big data [15]. If the reduction of data represents the removal of waste in data, it is indeed in line with the philosophy of lean management. Subsequently, the reduced big data can be transmitted to cloud-based services, which have strong data storage and analysis capabilities and can handle big data on behalf of enterprises [14]. A system architecture for implementing lean data is illustrated in Fig. 5.1.

Lean data is helpful for many shop floor management activities. An obvious example is **predictive maintenance**, which decides whether to repair machines in advance based on the results of analysis using **machine learning** techniques [16]. In order to predict when a machine will fail, various sensors are used to monitor the machine's current level, temperature, noise level, etc. These data fluctuate continuously, resulting in a huge amount of data that are difficult to analyze. Applying the concept of lean data to pre-screen such data can significantly improve the effectiveness and efficiency of the subsequent data analysis.

Fig. 5.1 System architecture for implementing lean data

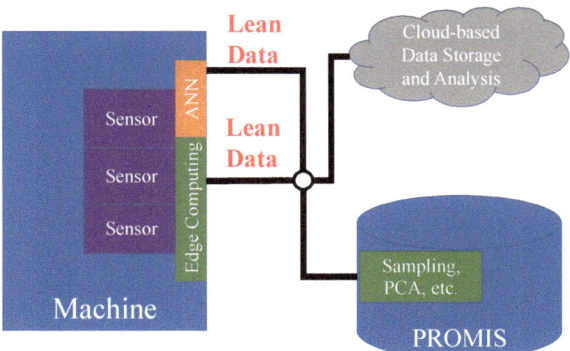

5.2.2 Lean Maintenance

The implementation of lean methods and tools in maintenance is called **lean maintenance** [17]. The goal of lean maintenance is to maintain the reliability of equipment at the lowest cost [18, 19].

A well-known concept in lean manufacturing is **TPM**, which is proactive and preventative maintenance to maximize the operational efficiency of equipment [8]. Here, operational efficiency means high equipment availability and few product defects. In other words, the goal of TPM is to simultaneously improve performance in many aspects, not just machine productivity.

The focus of TPM is on empowering operators to help maintain their own equipment, which will be much cheaper than asking the equipment supplier to do the maintenance (or repair), thus saving the associated costs. Therefore, TPM is a step towards lean maintenance.

In addition, **predictive maintenance** reduces unanticipated costly repairs caused by unexpected machine downs and is also a step towards lean maintenance [20]. Predictive maintenance is expected to evolve into **prescriptive maintenance**. According to the definition given by Limble CMMS [21], prescriptive maintenance is an extension of predictive maintenance. Based on the data collected by sensors, prescriptive maintenance applies built-in algorithms to analyze different scenarios (including production, maintenance, etc.), and then recommends the best scenario. An example comparing the difference between predictive maintenance and prescriptive maintenance is given below:

(Predictive maintenance)

The machine will be down in 17 h without maintenance. However, the next scheduled periodic maintenance is 30 h later. Therefore, the next periodic maintenance should be brought forwarded as a prevention measure.

(Prescriptive maintenance)

The machine will be down in 17 h without maintenance. Nevertheless, according to the analysis results using a fuzzy inference system, this time will be doubled if the temperature of the machine can be lowered to less than 45 °C. Therefore, the setting of the machine is adjusted according to this recommendation, and the next periodic maintenance will be carried out at the original time.

Further, **tele-maintenance** is a smart technology that enables equipment suppliers to perform maintenance remotely, rather than on site [22], thus saving transportation costs and coordinating schedules easily. Operators in the factory can also be instructed to maintain or repair equipment by themselves in the same way.

These maintenance policies have the following differences in reducing maintenance costs:

- Predictive maintenance is to reduce unexpected repair costs based on the results of data classification and forecasting (i.e., the application of AI).

Fig. 5.2 Lean maintenance and three existing maintenance policies

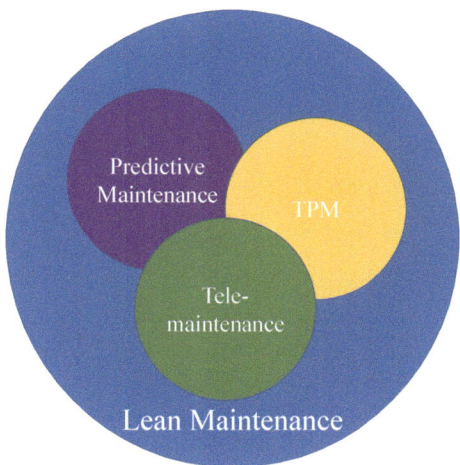

- TPM is to reduce the total costs of equipment maintenance and repair systematically (through personnel training).
- Tele-maintenance is driven by smart technologies (including both hardware and software) such as **5G**, **augmented reality (AR)**, etc.

The three maintenance policies overlap with each other, and all contribute to lean maintenance, as illustrated in Fig. 5.2. The intersection of the three maintenance policies is the following scenario:

A machine is equipped with various sensors. Then, based on the monitoring results, the operator of the machine performs preventive maintenance under the guidance of a remote equipment supplier.

5.2.3 Digitalized Kanbans

Traditional lean production systems control logistics and production with **withdrawal kanbans** and **production kanbans**, respectively. When the inventory in the buffer of a workstation (i.e., the supermarket) falls below a predetermined level, a withdrawal kanban will be released for the mover to pick up more workpieces from the previous workstation. One challenge is how to determine safety stock levels for workpieces in supermarkets. A low safety stock level can cause the next workstation to starve, while a high safety stock level can lead to inventory waste.

This function can be digitalized using smart technologies such as **radio frequency identification (RFID)** and **tablets** (or **smartphones**) as follows.

Step 1. First, a RFID tag is attached to the container of workpieces [23]. RFID is considered an **Internet of things (IoT)** technology [24].

Step 2. When the container is placed in the buffer, a nearby RFID receiver receives the emitted signal. The transmission distance of a RFID tag with a frequency of 16.56 MHZ is about 1.5 m.

Step 3. The RFID receiver is connected to a computer so that the number of workpieces in the buffer can be counted.

Step 4. When the number of workpieces in the buffer falls below the threshold, the computer sends a message (i.e., digitalized withdrawal kanban) to the mover's tablet (or smartphone).

Step 5. The mover picks up more workpieces from the previous workstation.

Production kanbans can be digitalized in a similar way. When the number of workpieces in the buffer falls below the threshold, the computer next to the buffer will send a message to the screen on the machine to display the type and number of workpieces to manufacture. After seeing this, the worker starts the operation.

Kanban digitalization saves the storage and management of kanbans, and also frees operators from thinking about how to use kanbans.

5.2.4 Machine Learning and Industry 4.0 Applications to Single-Minute Exchange of Die (SMED)

Single-minute exchange of die (SMED) is a lean philosophy to reduce machine setups (or changeovers) [7], since setup time represents a waste of capacity. For example, in the example discussed by Shingo and Dillon [25], the setup time of a job is fixed regardless of its size. As a result, as the job size increases, the setup time shared by each workpiece decreases, enabling SMED. However, the relationship between the total processing time of a job and its size may not be linear. Therefore, Carrizo Moreira et al. [7] suggested to minimize the time cost of a workpiece:

$$\text{Time cost} = \frac{\text{Total processing time} + \text{setup time}}{\text{Job size}} \tag{5.1}$$

However, large job sizes may not meet the ideal of one-piece flow in lean production [26].

In addition, if the production conditions of two jobs (or products) manufactured successively are similar, the time to re-setup can be shortened. However, this treatment may not be able to match the timing of customer demand.

Setup activities can be divided into **internal setup activities** and **external setup activities** [26]. Internal setup activities include changing machine dies (or molds) or fitting machines, which must be performed after a machine has been shut down. External setup activities, including preparing tools, can be performed while the machine is still running. Internal setup activities should be eliminated or converted into external setup activities [26]. Sometimes, it is not easy to tell whether a setup activity is internal or external. According to the view of Perico and Mattioli [6], machine learning can be applied to solve this problem. For example, a standard time must be established to regulate the time taken for a setup activity. To this end, an **ANN** can be constructed to decide how long a setup activity should take based on its properties (e.g., the number of setups per day, desired precision, the degree of expertise required, the degree of automation, etc.), as illustrated in Fig. 5.3. The training data are the standard times of setup activities that have been determined.

Example 5.1 To determine the standard time of a new setup operation, the data of ten similar setup operations have been collected for reference, as summarized in Table 5.1. An ANN with the configuration shown in Fig. 5.3 is constructed to predict the standard time of a setup operation a_j from its four attributes { x_{jk} }. Except the number of setups per day, the other attributes are based on experts' evaluations, and the evaluation results are integers within [1, 5]. The MATLAB program for implementing the ANN is shown in Fig. 5.4. All of the collected data are used to train the ANN, for which the training algorithm is the Levenberg–Marquardt (LM) algorithm [27]. The number of epochs is 10,000.

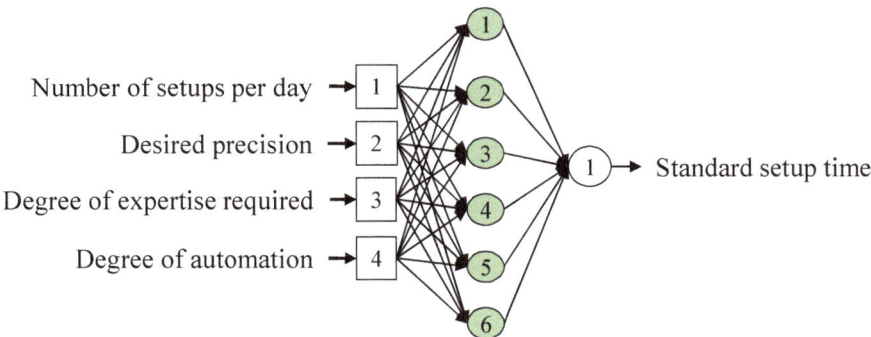

Fig. 5.3   ANN for establishing the standard time of a setup activity

Table 5.1   Data of some setup operations

j	x_{j1}	x_{j2}	x_{j3}	x_{j4}	a_j(min)
1	10	3	3	2	19
2	10	3	1	3	16
3	9	3	4	4	20
4	5	4	3	1	25
5	19	1	1	3	12
6	3	5	2	4	29
7	4	4	2	4	22
8	10	3	3	4	19
9	8	3	1	4	15
10	8	3	4	4	20

```
training_x=[10 10 9 5 19 3 4 10 8 8;3 3 3 4 1 5 4 3 3 3;3 1 4 3 1 2 2 3 1 4;2 3 4 1 3 4 4 4 4 4];
training_y=[19 16 20 25 12 29 22 19 15 20];
net=feedforwardnet(6);
net.dividefcn='dividetrain';
net.trainParam.lr=0.1;
net.trainParam.epochs=10000;
net.trainParam.goal=0.1;
net=train(net,training_x,training_y);
training_est_sst=net(training_x);
rmse=mean((training_y-training_est_sst).^2)^0.5;
```

Fig. 5.4 MATLAB code for implementing the ANN

The forecasting results are summarized in Fig. 5.5. The forecasting accuracy, measured in terms of RMSE, is 0.134 min. The values of the four attributes of the new setup operation are 9, 2, 4, and 3, respectively. After applying the trained ANN, as shown in Fig. 5.6, the standard setup time is established as 9.6 min.

Another AI technology that can also be applied to help establish the standard setup time from historical data is a classification and regression tree (CART) [28].

Example 5.2 In the previous example, a CART is constructed instead to establish the standard time of a new setup operation by referring to historical data. The MATLAB code for this purpose is shown in Fig. 5.7. The constructed CART is illustrated in

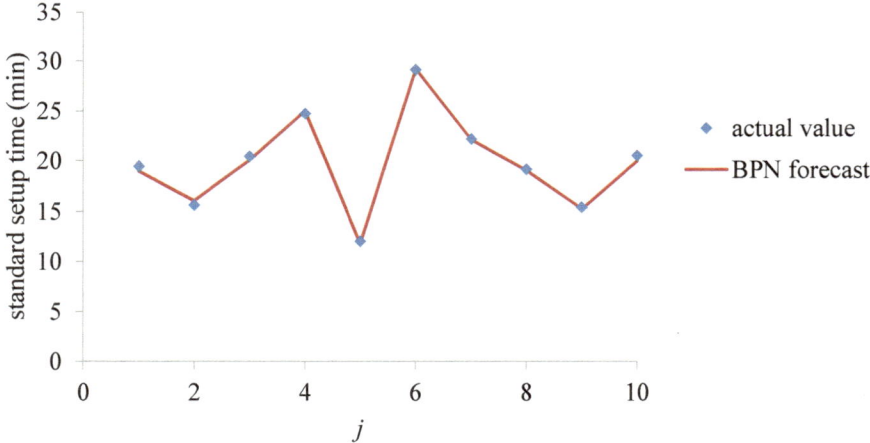

Fig. 5.5 Forecasting results

```
>> new_sst=net([9;2;4;3])

new_sst =

   9.4982
```

Fig. 5.6 Application of the trained ANN

```
y=[19;16;20;25;12;29;22;19;15;20];
sst_tree=fitrtree([10 3 3 2;10 3 1 3;9 3 4 4;5 4 3 1;19 0 1 3;3 5 2 4;4 4 2 4;10 3 3 4;8 3 1 4;8 3 4 4,],y);
view(sst_tree,'Mode','graph')
```

Fig. 5.7 MATLAB code for constructing the CART

Fig. 5.8 Built CART

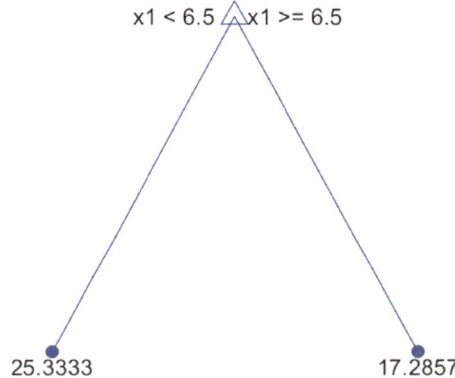

Fig. 5.8. The rules depicted in the CART are easy to interpret and communicate, which is the advantage of CART over ANN. However, the estimation accuracy may be low using CART. The standard setup time is established as 17.3 min.

A compromising approach is to express the rules derived using the ANN with a CART as follows:

- Simulating the attributes of many setup operations randomly.
- Applying the ANN to predict their standard setup times.
- Adjusting the format of the simulated data.
- Constructing a CART to parse the simulated data.

The MATLAB code for this purpose is shown in Fig. 5.9. The CART constructed to express the rules derived using the ANN is shown in Fig. 5.10.

The compromising approach provides an effective way to approximate an advanced AI technology with a basic AI technology, which is especially meaningful when AI technologies are to be introduced into a lean manufacturing system.

Perico and Mattioli [6] also mentioned a vision of applying **IoT** to **SMED**, in which a smart workpiece (a workpiece with smart tags) automatically transmits the setting such as the recipe and required tool to the next machine (a machine with signal receivers) before it reaches the machine. Then, the next machine starts to prepare automatically. When the smart workpiece arrives, the operation can start directly

```
random_x=[15 7 5 14 5 17 13 17 8 3 8 8 6 14 4 9 7 3 15 7 9 12 13 9 10 5 3 11 5 7 3 17 9 6 7 15 9 15
15 11 9 3 16 7 17 17 10 13 8 14;4 1 2 4 2 1 1 1 1 1 4 4 1 4 1 4 1 4 3 2 3 1 4 4 2 4 3 4 2 3 1 2 4 3 3 2 1
3 1 3 3 2 3 3 2 2 3 3 2 2;3 1 2 2 3 1 2 1 3 2 3 3 2 1 1 2 2 2 2 3 2 3 3 3 2 1 2 3 2 3 3 3 2 1 2 1 2 1 3 1 3 3 3
1 1 2 3 1 3 2 3 2;3 1 2 2 2 2 3 2 2 2 2 2 3 1 2 2 2 3 3 2 1 3 1 1 1 1 2 2 3 3 3 1 1 3 1 3 2 3 1 1 2 2 3 1 1
2 3 2 1 1];
random_y=net(random_x);
CART_x=transpose(random_x);
CART_y=transpose(random_y);
sst_tree2=fitrtree(CART_x,CART_y);
view(sst_tree2,'Mode','graph')
```

Fig. 5.9 MATLAB code for the compromise approach

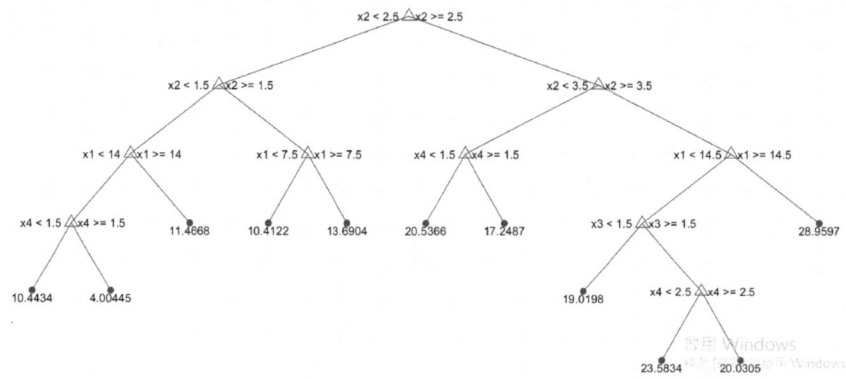

Fig. 5.10 CART for expressing the rules derived using the ANN

without waiting for setting, as illustrated in Fig. 5.11. In addition, a smart machine can monitor its own state and automatically generate performance reports [29].

Such an idea has so far been unrealistic. In addition, the black box-like approach of letting workpieces communicate with machines by themselves does not seem to be in line with the philosophy of lean manufacturing to make information, communication, and management transparent and humanized (i.e., understandable). Nevertheless, in our view, the point is not to introduce AI into lean manufacturing, but to combine the two to cope with the increasingly fierce industrial competition. To this end, there will be more and more AI black boxes in lean manufacturing. The applications of AI technologies must also be made increasingly transparent and easy to understand and communicate.

In addition, a more feasible approach derived from the above concept is called **predictive setup**: when an operator takes the tools required for an operation, he/she also takes the tools required for the next operation. Or, an operator notices a similarity between two consecutive settings, so there is no need to completely reset the previous setting.

Mass customization is a clear trend in many industries, so the importance of SMED will only increase with time.

Fig. 5.11 Application of
IoT to SMED

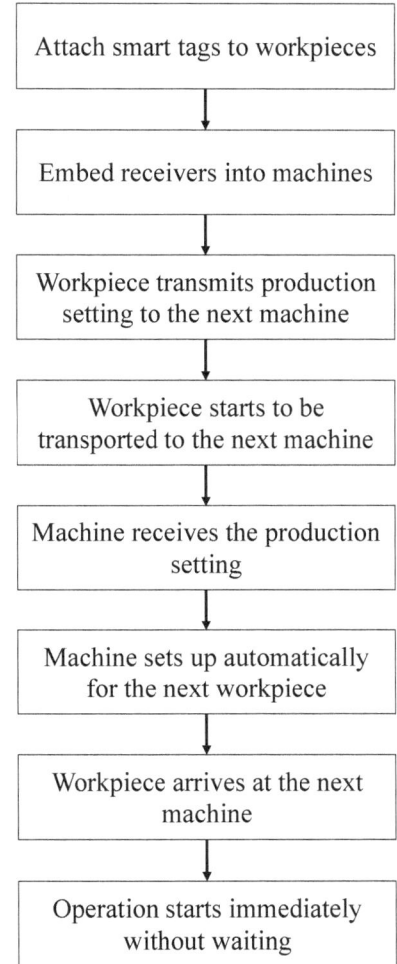

Attach smart tags to workpieces

Embed receivers into machines

Workpiece transmits production
setting to the next machine

Workpiece starts to be
transported to the next machine

Machine receives the production
setting

Machine sets up automatically
for the next workpiece

Workpiece arrives at the next
machine

Operation starts immediately
without waiting

5.2.5 Genetic Programming for Determining the Number of Kanbans

Determining the number of kanbans is a critical issue to a pull production system
[30]. Furthermore, dynamically changing the number of kanbans has been shown to
be an effective way to cope with demand fluctuation [31]. Belisário and Pierreval
[31] expressed the logic for adjusting the number of kanbans based on the subjective experience of production controllers in a manufacturing system with decision
trees. Then, they applied **genetic programming (GP)** to evolve these decision trees
to generate new decision trees (and rules) that correspond to better logic for the
same purpose. The effectiveness of new decision trees was evaluated by running a
production simulation of the manufacturing system.

GP is an evolutionary computing method in which programs used to do something are evolved to generate new programs that can do the same thing better. GP is an extension of genetic algorithms (GAs) [32]. In GP, programs, not genes, evolve. In theory, programs in GP can be written in any programming language. An example is given below.

Example 5.3 According to the subjective experience of a production control staff, the logic shown in Fig. 5.12 is currently followed to adjust the number of kanbans in the manufacturing system. The logic is expressed as a decision tree in Fig. 5.13. The decision tree is read from the lower left node to the upper right.

Example 5.4 Two decision trees (i.e. parent decision trees) for adjusting the number of kanbans in the manufacturing system are shown in Fig. 5.14. A crossover operation is performed to combine the two decision trees to generate new decision trees. The results (i.e., child decision trees) are shown in Fig. 5.15. The crossover operator works by swapping arbitrary branches of the two decision trees (indicated by gray blocks).

Example 5.5 The logic of each decision tree is evaluated by applying it to adjust the number of kanbans in the manufacturing system. Then, the fitness of a chromosome

```
If work-in-process (WIP) ≥ 100 Then
   Number of kanbans increases by 1
Else
   If WIP < 50 Then
      Number of kanbans decreases by 1
   Else
      If factory utilization (FU) ≥ 95% Then
         Number of kanbans increases by 1
      End If
   End If
End If
```

Fig. 5.12 Current logic for adjusting the number of kanbans

Fig. 5.13 Decision tree of the logic

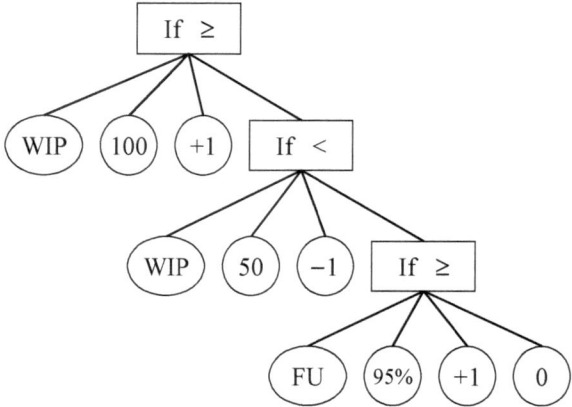

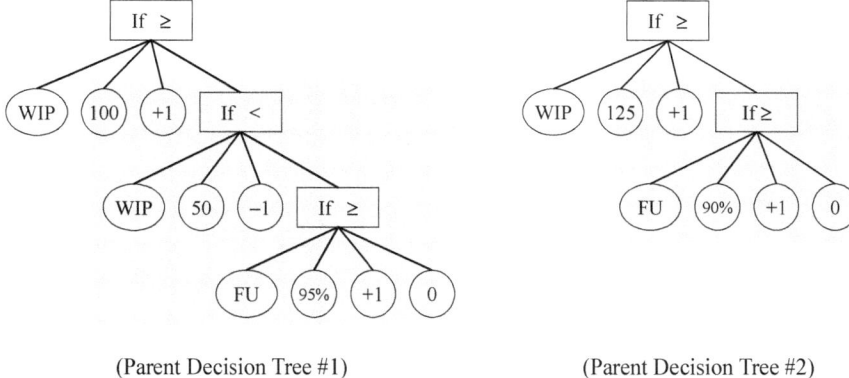

(Parent Decision Tree #1)   (Parent Decision Tree #2)

Fig. 5.14 Two parent decision trees

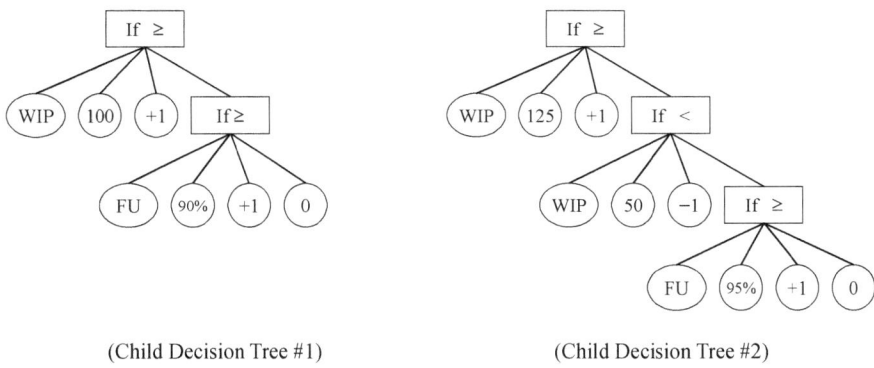

(Child Decision Tree #1)   (Child Decision Tree #2)

Fig. 5.15 Child decision trees

(i.e., decision tree) is measured in terms of the average cycle time. The results are summarized in Table 5.2. Obviously, the fitness has improved after evolution.

In GP, mutation is implemented by randomly removing a subtree at a selected point and replacing it with a randomly generated subtree. However, the mutation operator is seldom used (Fig. 5.16).

Table 5.2 Fitness achieved by each decision tree	Decision tree	Fitness (hr)
	Parent decision tree #1	36
	Parent decision tree #2	42
	Child decision tree #1	39
	Child decision tree #2	33

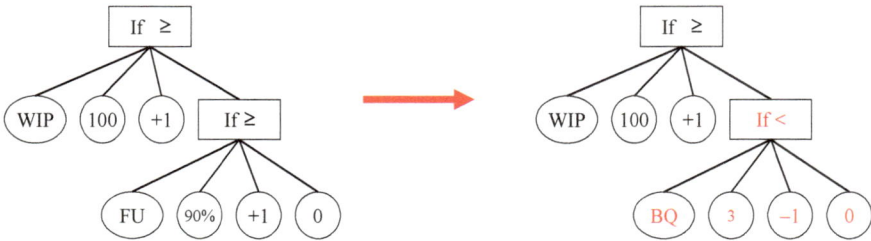

Fig. 5.16 Mutation

5.3 Conclusions

From the discussion in the above chapters, the following conclusions can be drawn, that is, the combination of lean manufacturing and AI is not only a way to cope with the increasingly fierce industrial competition but also a feasible approach. The combination of the two has the following possibilities:

- The introduction of AI technology applications into lean manufacturing systems: In order to comply with the transparent management philosophy of lean manufacturing systems, the applied AI technologies must be basic, mature, and easy to learn, understand, and communicate.
- An alternative is to approximate an advance AI technology with a basic AI technology, as illustrated in Example 5.2.
- The applications of more lean manufacturing techniques or concepts to general manufacturing systems that are highly AI-enabled: Many lean manufacturing concepts have been widely used in various types of manufacturing systems. These manufacturing systems are subject to their own internal and external restrictions, and it is difficult to fully promote the philosophy of lean manufacturing in them. However, trying to make the AI technology applications in these manufacturing systems more transparent can also be in line with the philosophy of lean manufacturing.

References

1. SESA Systems, Shop floor management, a visual management approach to the shop floor (2021). https://www.sesa-systems.com/en/shop-floor-management
2. Oracle.com, Understanding shop floor management (2021). https://docs.oracle.com/en/applications/jd-edwards/supply-chain-manufacturing/9.2/eoash/understanding-shop-floor-management.html#understanding-shop-floor-management
3. R.B. Freeman, M.M. Kleiner, The last American shoe manufacturers: decreasing productivity and increasing profits in the shift from piece rates to continuous flow production. Ind. Relat. J. Econ. Soc. **44**(2), 307–330 (2005)
4. Lean Enterprise Institute, Pull production (2021). https://www.lean.org/lexicon-terms/pull-production/

5. X.A. Koufteros, Testing a model of pull production: a paradigm for manufacturing research using structural equation modeling. J. Oper. Manag. **17**(4), 467–488 (1999)
6. P. Perico, J. Mattioli, Empowering process and control in lean 4.0 with artificial intelligence, in *Third International Conference on Artificial Intelligence for Industries* (2020), pp. 6–9
7. A. Carrizo Moreira, G. Campos Silva Pais, Single minute exchange of die: a case study implementation. J. Technol. Manage. Innov. **6**(1), 129–146 (2011)
8. leanproduction.com, TPM (total productive maintenance) (2021). https://www.leanproduction.com/tpm/
9. J.J. Dahlgaard, G.K. Khanji, K. Kristensen, *Fundamentals of Total Quality Management* (Routledge, 2008)
10. leanproduction.com, Total quality management (2021). https://www.lean.org/lexicon-terms/total-quality-management/
11. J. Nandimath, E. Banerjee, A. Patil, P. Kakade, S. Vaidya, D. Chaturvedi, Big data analysis using Apache Hadoop, in *IEEE 14th International Conference on Information Reuse & Integration* (2013), pp. 700–703
12. A. Navlani, Neural network models in R (2019). https://www.datacamp.com/community/tutorials/neural-network-models-r
13. T. Küfner, T.H.J. Uhlemann, B. Ziegler, Lean data in manufacturing systems: using artificial intelligence for decentralized data reduction and information extraction. Procedia CIRP **72**, 219–224 (2018)
14. T. Hafeez, L. Xu, G. Mcardle, Edge intelligence for data handling and predictive maintenance in IIOT. IEEE Access **9**, 49355–49371 (2021)
15. C.S. Liew, A. Abbas, P.P. Jayaraman, T.Y. Wah, S.U. Khan, Big data reduction methods: a survey. Data Sci. Eng. **1**(4), 265–284 (2016)
16. T.P. Carvalho, F.A. Soares, R. Vita, R.D.P. Francisco, J.P. Basto, S.G. Alcalá, A systematic literature review of machine learning methods applied to predictive maintenance. Comput. Ind. Eng. **137**, 106024 (2019)
17. H.M. Hashemian, State-of-the-art predictive maintenance techniques. IEEE Trans. Instrum. Meas. **60**(1), 226–236 (2010)
18. E. Ramos, R. Mesia, C. Alva, R. Miyashiro, Applying lean maintenance to optimize manufacturing processes in the supply chain: A Peruvian print company case. Int. J. Supply Chain Manage. **9**(1), 264–281 (2020)
19. K. Antosz, L. Pasko, A. Gola, The use of artificial intelligence methods to assess the effectiveness of lean maintenance concept implementation in manufacturing enterprises. Appl. Sci. **10**(21), 7922 (2020)
20. S. Selcuk, Predictive maintenance, its implementation and latest trends. Proc. Instit. Mech. Eng. Part B: J. Eng. Manuf. **231**(9), 1670–1679 (2017)
21. Limble CMMS, A complete guide to prescriptive maintenance (2022). https://limblecmms.com/blog/prescriptive-maintenance/
22. E. Garcia, H. Guyennet, J.C. Lapayre, N. Zerhouni, A new industrial cooperative telemaintenance platform. Comput. Ind. Eng. **46**(4), 851–864 (2004)
23. X. Shi, D. Tao, S. Voß, RFID technology and its application to port-based container logistics. J. Organ. Comput. Electron. Commer. **21**(4), 332–347 (2011)
24. T. Chen, Y.C. Wang, An advanced IoT system for assisting ubiquitous manufacturing with 3D printing. Int. J. Adv. Manuf. Technol. **103**(5), 1721–1733 (2019)
25. S. Shingo, A.P. Dillon. *A Revolution in Manufacturing: The SMED System* (Routledge, 2019)
26. J. Miltenburg, One-piece flow manufacturing on U-shaped production lines: a tutorial. IIE Trans. **33**(4), 303–321 (2001)
27. G. Lera, M. Pinzolas, Neighborhood based Levenberg-Marquardt algorithm for neural network training. IEEE Trans. Neural Netw. **13**(5), 1200–1203 (2002)
28. C.D. Sutton, Classification and regression trees, bagging, and boosting. Handbook Statist. **24**, 303–329 (2005)

29. G.Y. Lee, M. Kim, Y.J. Quan, M.S. Kim, T.J.Y. Kim, H.S. Yoon, S. Min, D.-H. Kim, J.-W. Mun, J.W. Oh, I.G. Choi, C.-S. Kim, W.-S. Chu, J. Yang, B. Bhandari, C.-M. Lee, J.-B. Ihn, S.H. Ahn, Machine health management in smart factory: a review. J. Mech. Sci. Technol. **32**(3), 987–1009 (2018)

30. H.C. Co, M. Sharafali, Overplanning factor in Toyota's formula for computing the number of kanban. IIE Trans. **29**(5), 409–415 (1997)

31. L.S. Belisário, H. Pierreval, Using genetic programming and simulation to learn how to dynamically adapt the number of cards in reactive pull systems. Expert Syst. Appl. **42**(6), 3129–3141 (2015)

32. P.G. Espejo, S. Ventura, F. Herrera, A survey on the application of genetic programming to classification. IEEE Trans. Syst. Man Cybern. Part C (Applications and Reviews) **40**(2), 121–144 (2009)